WEATHER

WEATHER

STORM DUNLOP

THUNDER BAY
P·R·E·S·S

San Diego, California

THE EARTH'S CLIMATE IS A KEY PROTAGONIST IN THE STORY OF HUMANITY: our survival as a species has depended on cycles of freeze and thaw, and the ability of our ancestors to adapt to changes in temperature and terrain. The warming and cooling of the planet's oceans and land is vital to its ability to support life. Humans and wildlife migrate from continent to continent and inhabit different parts of the globe, seeking food and shelter at the mercy of the weather. As this book shows, life and climate go hand-in-hand, and one of my challenges as a climber is to anticipate changes in the weather and experience the very limits of this partnership. Choosing to press on or retreat in time, before the arrival of a severe storm, can be a life-or-death decision. The weather not only brings its wind and storms but also affects the nature of the snow cover and the risk of avalanche or stone fall.

The changing weather of our planet—which has experienced ice ages and droughts throughout its history—has of course physically shaped the landscape, as well as the life it supports, and the superb images in this book are testament to this. From the wind-sculpted rocks, or yardangs, in the deserts of Asia, to the vast forests of the subarctic taiga, *Weather* presents the evidence of the continuing cycles of winds, rain, cloud, and atmospheric optics everywhere in the world. From the immense scale of vast glaciers, a reminder of our once-frozen past, as well as a stark symbol of our precarious future, to the intricacy of beautiful eroded wadis in the Arabian Desert, which seen from above look like tiny rivulets trickling off the landscape, our planet's unique variety of climates is both surprising and spectacular. Of course, dramatic one-off events, such as storms, floods, and hurricanes, also bring chaos and devastation in their wake, leaving their own mark on the landscape as well as on history.

Subtle and sometimes stunning visual manifestations of weather, on the other hand, such as auroras and rainbows, provide inspiration for poets but rarely make front-page news. Long ago thought of as a sign from a powerful deity, the aurora, is, in fact, a dramatic phenomenon brought

FOREWORD

about by charged particles from the Sun entering the upper atmosphere. Similarly, halos are created by the complex interplay of light reflection and refraction, and while these can be explained, and Storm Dunlop does so succinctly and clearly, still one cannot help but wonder at the power of nature and the apparent intelligence behind such beauty.

Dependent as we are on the elements, so too, we increasingly realize, is the climate affected by humans. In fact, some of those areas that we have prized for their favorable climates, and subsequently populated, are experiencing all too clearly the devastating effects of industry, fuel-dependency, and general human over-consumption on our weather patterns. Recent reports suggest that hurricanes in the Gulf of Mexico and surrounding areas are on the increase due to global warming, caused by the burning of fossil fuels, which has the effect of warming our seas. Equally, parts of the Southern Hemisphere are feeling the effects of depleting ozone in the upper levels of the atmosphere. Throughout all these areas, the precious balance of the world's climates looks increasingly precarious.

There is much that we can predict, and the images inside—including sophisticated satellite imagery—are proof of the leaps we have made in even a few years in forecasting the weather. While we do know that our climate is changing, the powerful images here offer a comprehensive picture of not only how far we have come in our understanding of weather but also how far we still have to go.

AT FIRST SIGHT, THE INFINITE VARIETY OF CLOUDS APPEARS TO DEFY CLASSIFICATION, but in fact, there are relatively few—ten—major forms (known as "genera"), and these are usually easy to distinguish. Although the general nature of clouds has been known for many centuries, it was not until 1802 that the British pharmacist, Luke Howard, introduced a classification scheme similar to that used for plants and animals. Although modified extensively since that time, some of Howard's terms remain in use today.

Meteorologists divide clouds into two broad types: cumulus, or heaped clouds; and stratus, or layer clouds. Cumuliform clouds are usually associated with instability, the condition where a pocket of air, rising in the atmosphere, perhaps as a thermal over warmed ground, remains warmer than its surroundings and thus continues to rise. By contrast, stratiform clouds are normally associated with stability. If the air in a particular layer is displaced—perhaps being forced upward by the wind passing over a mountain—when conditions are stable, the air sinks back down toward its original level after passing the obstacle.

Somewhat confusingly to the novice, some clouds exhibit both cumuliform and stratiform characteristics. Stratocumulus and altocumulus, for example, often occur as a layer of cloud that has been broken up into individual cloudlets by shallow convection. An additional group is sometimes mentioned: cirriform clouds. These are high clouds that consist of ice crystals and thus display a distinct fibrous appearance. Cloud types may also be subdivided according to their heights. There are three ranges (known as *étages*): low, medium, and high. Low clouds are cumulus, stratus, and stratocumulus; medium: altocumulus, altostratus, and nimbostratus. (The last may sometimes descend to the surface.) High clouds are cirrus, cirrostratus, and cirrocumulus. Cumulonimbus may extend through all three levels. Actual heights vary considerably depending on latitude. As a rule, their bases and tops are lower near the poles. It is notable that although meteorologists worldwide use metric units, cloud heights are still specified in feet, because this unit has been adopted internationally by the airline industry.

The tops of giant cumulonimbus clouds in the tropics may reach 60,000 ft (approx. 18 km) or more, whereas toward the poles only very rarely do clouds reach 26,000 ft (approx. 8 km) and are generally much lower. The base of low clouds, by contrast, is relatively constant at about 6,500 ft (approx. 2 km), although it may sometimes descend as low as the surface.

The classification scheme has further subdivisions: species, which describe general shape and structure, and varieties, which cover cloud transparency and the overall arrangement of individual cloud elements. Most are not shown in detail here, but a few supplementary features have been included because of their striking nature. There are also some clouds that occur far higher in the atmosphere: the rare but outstandingly beautiful displays of nacreous clouds, which are so striking that they are usually mentioned on the national news, and noctilucent clouds, the highest of all, glowing in the sky at midnight.

1 | BLUE SKIES BECOMING OVERCAST

Nacreous, or mother-of-pearl clouds

THESE POLAR STRATOSPHERIC CLOUDS occur at altitudes of 9–19 miles/15–30 km in the lower stratosphere, where normally clouds are rare. Their structure is complex, with initial condensation nuclei of sulfuric acid droplets, on which nitric acid trihydrate is deposited when temperatures drop to -108.4°F/-78°C. The tiny particles (about 1 μm in diameter) are hard to detect, but when they enter a colder region—perhaps as a result of wave motion of the air created by mountains far below—with a rapid drop in temperature below -117.4°F/-83°C, they become coated with a layer of ice. Diffraction of sunlight by the ice-covered particles—still only about 2 μm in diameter—gives rise to pure spectral colors. Slower cooling, as at the onset of winter, produces fewer, but larger particles (about 10 μm), giving rise to clouds that appear white.

Previous pages | A lenticular cloud

A LENTICULAR CLOUD (ALTOCUMULUS LENTICULARIS), photographed with a fish-eye lens over Sandstone Bridge, Alabama Hills, California, at sunset. Lenticular clouds (also known as wave clouds) are produced by the motion of the air above hills or mountains, and may persist for a long time if the wind remains constant in strength and direction. Sandstone Bridge is a natural arch formed by water erosion.

Nacreous clouds

SEEN JUST BEFORE SUNRISE or after sunset, these beautiful clouds hold a hidden hazard. Their tiny particles are a significant factor in the destruction of stratospheric ozone and the creation of ozone holes. Chlorine from chlorofluorocarbons (CFCs) in the atmosphere becomes active on their surfaces, breaking down ozone. In addition, they remove nitrogen from the air, which would otherwise combine with chlorine to give inert chlorine nitrate. These clouds have become more frequent in recent years, but the reasons for this are unknown.

Previous pages | Noctilucent clouds

THESE SILVERY-WHITE OR SLIGHTLY YELLOWISH CLOUDS are the highest in the atmosphere, occurring at 50–53 miles/80–85 km, in the mesosphere. They are visible around midnight for a period of about two months around the summer solstice, at high latitude (above about 45° N or S), when the clouds are illuminated by sunlight but the observer is in darkness. They consist of ice particles, formed on freezing nuclei: perhaps consisting of meteoric dust or clusters of ions produced by cosmic rays. The clouds form a thin sheet, and their billows and other forms are produced by waves in the upper atmosphere. The patches of cloud often move west, while the waves themselves travel in the opposite direction. Noctilucent clouds seem to be becoming more frequent, the reasons for which are not known.

Cumulus clouds

CUMULUS CLOUDS ARE ONE OF A GROUP OF LOW CLOUDS, having bases below about 6,500 ft (approx. 2 km). This photograph, taken in Western Australia, shows how cumulus are separated from one another by distances that are roughly as large as the clouds themselves. Depending on the conditions, such an arrangement may persist all day, with individual clouds growing and then dispersing, or the clouds may spread out into a layer of stratocumulus, with narrow gaps between the individual pancakelike cloudlets. With strong solar heating, however, large cumulonimbus clouds are likely to form instead.

Cumulus clouds at sunset

THESE "FAIR-WEATHER CUMULUS" speak of the end of a fine day. Cumulus clouds typically have rounded tops and flat bases. They are frequently produced by localized convective cells ("thermals") that break away from surfaces that have been heated by the Sun and rise until they reach the condensation level—marked by the flat bases. If the heating is particularly strong, they may go on to build much deeper clouds, towering cumulus, or even cumulonimbus clouds. Small cumulus (as here) decay and fade away when solar heating dies down at sunset.

Medium-sized cumulus

MEDIUM-SIZED CUMULUS often show distinct, rounded tops and indications that the convective cells within them are still growing vertically. The cells are leaning slightly toward the right, indicating that the wind is from the left. (Wind speeds generally increase with height, so it is common to see cumulus towers leaning in this way.)

Heavy cumulus clouds

THESE SUBSTANTIAL CLOUDS are the largest form of cumulus (cumulus congestus). The main convective cells still have sharply outlined, rounded tops, which is an indication that freezing has not yet started in the highest regions. Once freezing (known technically as glaciation) begins, the cloud tops become less distinct and "softer." The clouds would then be classed as cumulonimbus. The bases of these clouds are very dark, and it is possible that they may shortly give rise to precipitation. Rain from cumulus congestus is common in the tropics and also occurs in temperate regions during summer.

Previous pages | Layer clouds from above

TAKEN AT THE TOP OF MAUNA KEA, HAWAII, this photograph shows two types of layer (stratiform) clouds. Close up and in the distance are banks of deep nimbostratus clouds, and between them an indistinct, hazy layer of altostratus. Seen from below, the nimbostratus would appear as a heavy dark cloud, producing considerable rainfall, and reaching down fairly close to the ground. The base of the altostratus would be higher and, depending on its thickness, the outline of the Sun might just be visible through the cloud, appearing as if through ground glass. A site of a major astronomical observatory, Mauna Kea has prevailing clear, stable conditions, well above the cloud.

Cirrus clouds over a sand dune

THESE CIRRUS CLOUDS OVER KANSAS are less organized than those in the image at right, above, with apparently random swirls of cloud. They are a general indication of fairly quiet conditions, but there are suggestions toward the horizon that they are becoming thicker and covering a greater area of the sky. A low pressure area may be approaching from that direction.

Cirrus ahead of a depression

CIRRUS (ABOVE) ARE THE HIGHEST CLOUDS commonly seen in the atmosphere and generally have bases at 20,000 ft (approx. 6 km), although they may be only half that height at very high latitudes. They consist of ice particles. Long trails ("mare's tails"), as here, indicate strong wind shear at height. When they gradually cover more and more of the sky, all advancing from one direction, they are generally a reliable indication of an approaching depression.

Cirrus floccus

THESE CIRRUS CLOUD TUFTS (BELOW) are seen almost vertically overhead. They show no sign of being drawn out into long streaks and occur as individual tufts (known technically as cirrus floccus). They suggest that the weather is fair and likely to continue so for a while, although the numerous clumps of cloud indicate that there is some instability at height, and this could become significant and assist the growth of cumulonimbus clouds if the latter build up to their height.

Cirrostratus clouds and contrails

THE SKY IS COVERED WITH A VEIL of cirrostratus cloud, thickening ahead of the warm front of an approaching depression. When the cirrostratus was thinner, halo phenomena (see page 146) were almost certainly visible. The four persistent contrails (condensation trails) from aircraft are a sign of increasing humidity at height. When the air at altitude is dry, any contrails disperse rapidly. Ahead of a depression, however, they last for a long time and have even been known to completely blanket the sky.

Cirrocumulus billows

LINES OF SMALL, CIRROCUMULUS CLOUDLETS fill the sky. The pattern is commonly known as a "mackerel sky"—although this term is also frequently applied to altocumulus with a similar appearance. Like the other cirrus clouds, cirrocumuli mainly consist of ice particles. The high and thin cloudlets show no signs of shading, as distinct from lower altocumulus clouds, which are thicker with shading on the shadowed sides. The billows are an indication of wind shear at altitude, and the billows here lie across the wind. Small patches of cirrocumulus have little significance, but when the whole sky is covered, as here, it usually implies a depression is approaching.

BLUE SKIES BECOMING OVERCAST

Stratocumulus clouds over the Pacific

THESE CLOUDS MARK THE BOUNDARY of a mass of warm air (top left) that has come from the Californian coast some 360 miles/580 km away. Usually, the clouds form just off the coast. In this instance, however, hot Santa Ana winds from the continental interior have blown the warm air much farther over the ocean before the clouds have formed. To meteorologists' surprise, early satellite images revealed that stratocumuli are the most common clouds, occurring over large areas of the oceans.

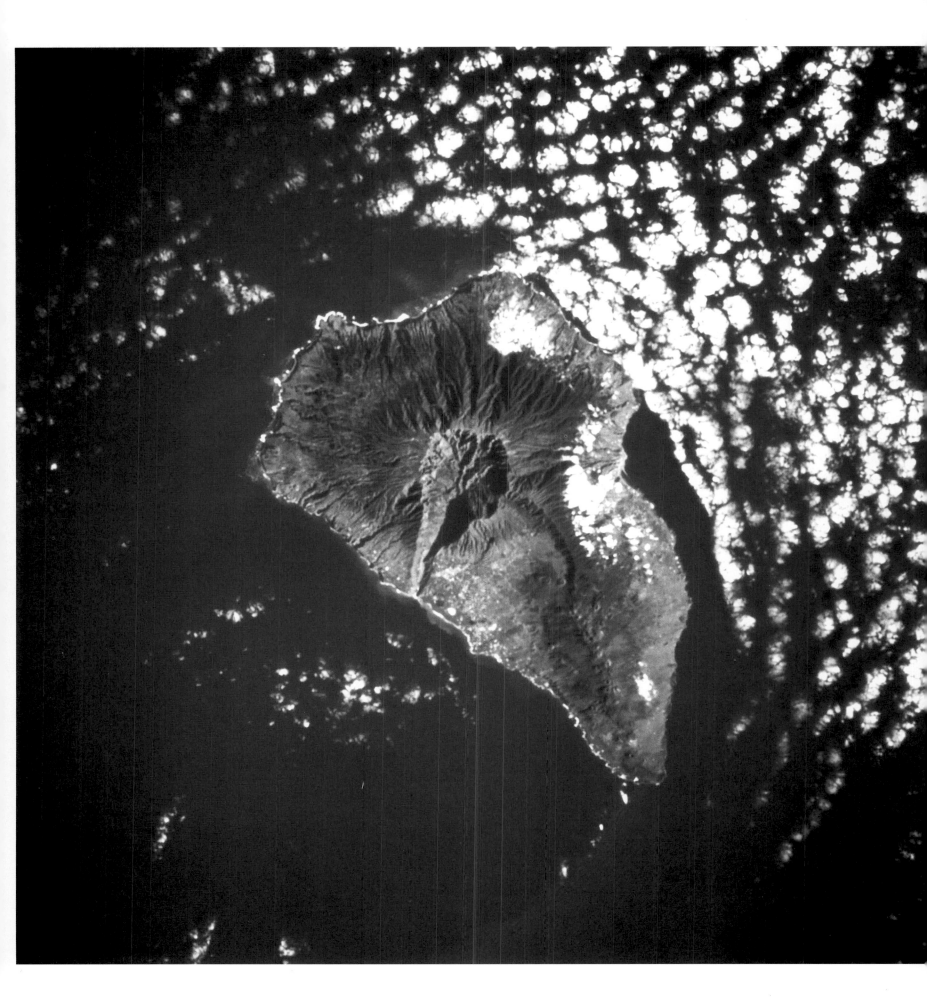

Stratocumulus clouds near La Palma

STRATOCUMULUS—LARGE "PANCAKES" OF CLOUD with narrow clear spaces between them—lie immediately east of La Palma, the largest of the western Canary Islands, in the Atlantic Ocean. North is at top left. The volcanic caldera is clearly visible, and this one is the original feature to have been given that name. ("Caldera" means "cauldron" in Spanish.) It has been suggested that the western half of the island is unstable and could collapse into the sea, giving rise to a massive tsunami that would swamp Caribbean islands and devastate the eastern coast of the United States, traveling up to 12 miles/20 km inland.

ALTOCUMULUS ARE MIDDLE-LEVEL CLOUDS with bases anywhere between 6,500 and 20,000 ft (approx. 2–6 km), but generally below 12,000 ft (approx. 4 km) at high latitudes. They show distinct shading (unlike the higher cirrocumulus) and frequently accompany lower clouds, such as the cumulus seen here. Mount Erebus on Ross Island in Antarctica is the southernmost active volcano in the world.

Altocumulus clouds from below

THE ALTOCUMULUS CLOSELY RESEMBLES THE LOWER STRATOCUMULUS, with "pancakes" of cloud, separated by narrow gaps of clear air. The two forms are distinguished by their size: altocumulus cloudlets are always less than 5° across, whereas the elements of stratocumulus are larger. Altocumulus may consist of water droplets or a mixture of droplets and ice crystals. Under cold conditions they may give rise to snow, although the particles frequently melt and turn to fine rain before they reach the ground.

Altocumulus billows

BILLOWS SUCH AS THESE (KNOWN AS "MACKEREL SKY") FORM quite frequently when there is wind shear at height, with one layer of air moving faster or in a slightly different direction from the layer below it. Altocumulus clouds are defined as having individual cloudlets that are smaller than 5° across—unlike the larger stratocumulus—and larger than 1° across—below which they would be classed as cirrocumulus. Here the clouds merge toward the horizon into a layer of altostratus cloud.

Mature cumulonimbus incus

CUMULONIMBUS CLOUDS FREQUENTLY RISE until they reach the tropopause, an inversion (a region where temperature increases with height). This marks the top of the lowest layer (the troposphere) of the atmosphere. Upward cloud growth is checked, they spread out into a characteristic anvil-shaped top, and they are then known as cumulonimbus incus. The bases may be at the lowest level, but their tops may be as high as 60,000 feet (about 18 km) in the tropics. Here, two individual cells have become glaciated and spread out into anvils. Also visible is the heavy precipitation that is typical of cumulonimbus clouds.

Cumulonimbus cloud with pileus

THIS RAPIDLY GROWING CLOUD has just reached the stage when a cumulus congestus becomes cumulonimbus with the formation of ice crystals in the topmost layers. Two of the active cloud towers have lifted a humid layer of air above the condensation and freezing level, creating a veil of cloud draped over them. This form, known as pileus, is only temporary, because the rising cloud towers will soon penetrate the pileus layer, which will then be incorporated into the active cells by the circulation existing around them.

Cumulonimbus clouds at sunset

THIS CLUSTER OF CUMULONIMBUS CELLS over the sea illustrates how such systems develop. Three convective cells ("thermals") have reached the inversion and spread out to form anvils, with the oldest the farthest away. In front of them lies what is known as a "flanking line" of additional, growing cells, which gradually increase in height toward the body of the storm. If the sea is warmer than the air above it, the cells will continue to grow, even though solar heating has ceased with nightfall.

Cumulonimbus with anvil approaching

CUMULONIMBUS CLOUDS MAY PRODUCE heavy rain and hail, and turn into thunderstorms, regardless of whether they have the typical anvil top. They are the "showers" frequently described in weather forecasts. The largest clusters of active cells may become multicell storms or even supercell storms, which may persist for many hours and produce violent weather, including spawning highly destructive tornadoes.

Cumulonimbus clouds seen from space

THIS PHOTOGRAPH, TAKEN FROM LOW EARTH ORBIT, shows cumulonimbus clouds in various stages of growth, with two having developed anvils. From this viewpoint it is easy to see the "overshooting tops," where the extremely vigorous updrafts have penetrated some distance above the inversion at the top of the clouds. The haze in the background of this photograph of Zaire was caused by the burning of agricultural land.

Orographic cloud, Greenland

THIS PHOTOGRAPH SHOWS CLOUD forming where the wind is forcing moist air from the sea to rise over the mountain. In so doing, it reaches the condensation level and produces a layer of cloud, which dissipates to leeward of the mountain where the air descends again. Although the cloud is predominantly stratiform, there are some slight signs of instability in the rounded heads. In the dry Arctic desert, here on Clavering Island in northeast Greenland, rainfall from such clouds is extremely significant and helps to support the hardy vegetation on the mountain slopes.

Orographic cumulus cloud

THESE CLOUDS ARE BUILDING UP in the early afternoon over the southern slopes of Mount Piccolo Altare in the Ligurian Apennines in Italy. The uplift provided by the hills often engenders considerable instability, leading to massive cumulonimbus clouds and accompanying heavy showers. In extreme cases, orographic lifting has led to intense thunderstorms that remain stationary over the peaks and produce destructive flash flooding.

Von Karman vortices in stratocumulus clouds

THESE CLOUD VORTICES IN A SHEET OF STRATOCUMULUS have been created downwind of the peak of Alexander Selkirk Island (5,249 ft/1,600 m), which has disrupted the steady flow. Known as von Karman vortices, this phenomenon occurs at all levels and is frequently seen associated with mountainous islands such as Svalbard in the Arctic and the Canary Islands in the North Atlantic. Alexander Selkirk Island and Robinson Crusoe Island are part of the Juan Fernandez Islands in the southern Pacific Ocean off the coast of Chile.

Previous pages | Mamma beneath cumulonimbus anvil

THESE DRAMATIC BULGES FREQUENTLY DEVELOP beneath the overhanging anvils of large cumulonimbus clouds. The top of the anvil radiates heat upward, cooling the cloud, and cold air sinks beneath it in a form of "upside-down convection." Often mamma take the form of long, irregular tubes, rather than the orderly bulges seen here. Similar features are found in a number of other cloud types including cirrus, cirrocumulus, altocumulus, and stratocumulus.

Mamma and crepuscular rays

THE MAMMA IN THIS PHOTOGRAPH are somewhat unusual, in that they have developed beneath a layer of cloud. One portion of the cloud, where the mamma have formed, appears to be stratocumulus, but other portions are relatively featureless stratus. Because of the low angle of illumination, with sunlight filtering in from the edge of the cloud sheet, the mamma are casting distinct shadows (crepuscular rays) on the underside of the cloud.

Mamma

THE MAMMA HERE SHOW THE TYPICAL appearance of those that develop beneath cumulonimbus 43
clouds. In this example, the mamma are shades of gray since dramatic lighting from low sunlight
is absent. Despite sometimes seeming turbulent and threatening, such formations do not
generally herald any dramatic events. They tend to develop at the rear of thunderstorms and,
although they sometimes occur when tornadoes develop, cannot be taken as an indication that
such an event is imminent.

"Une pile d'assiettes"

FREQUENTLY, LENTICULAR CLOUDS FORM as a stack of clouds, one above the other, and this 45
appearance is known as *une pile d'assiettes* (French for "a pile of plates"). They indicate that
there is a series of humid layers, separated by drier layers that are all forced to rise and fall in the
waves. Depending on the wind strength and direction, and also the nature of the obstacle, more
than one stack may occur in the train of waves to leeward of the peaks. These clouds were
photographed over the island of South Georgia, in the South Atlantic Ocean.

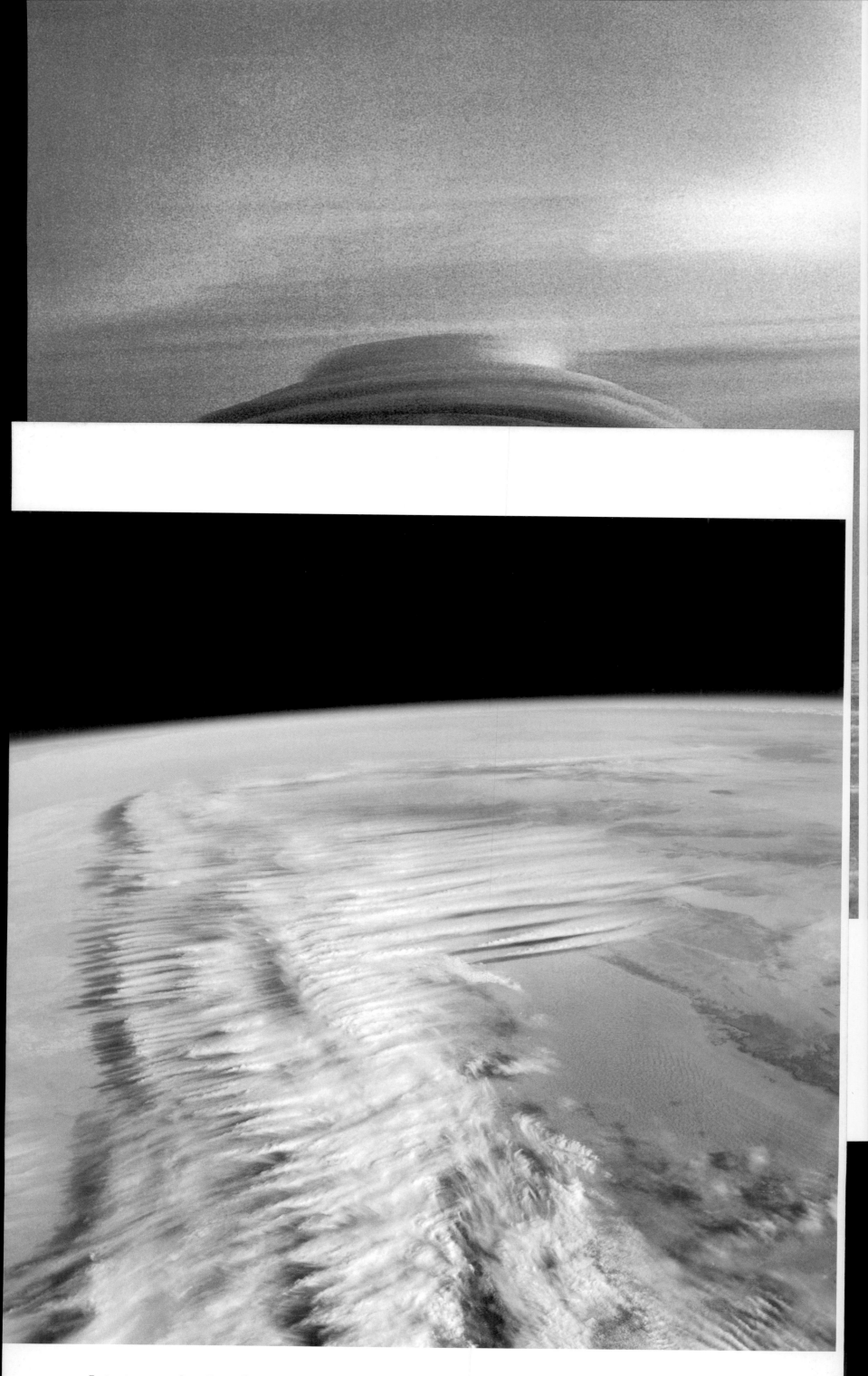

Jet-stream clouds, Libya

JET STREAMS ARE HIGH-ALTITUDE, HIGH-SPEED RIBBONS of air that develop where there are extreme temperature contrasts in the atmosphere. They may be thousands of kilometers long, hundreds of kilometers wide, and just a few kilometers deep. Cirrus clouds frequently form within them and betray their presence, although at other times the jet streams may be invisible. The clouds often form as billows across the flow or as long streaks along the stream.

Jet-stream clouds, eastern Canada

FOR A FLOW TO BE CALLED A JET STREAM, the wind speed must exceed 56–62 mph/90–100 kph. The maximum speed ever recorded was 408 mph/656 kph in a jet stream observed over South Uist in the Scottish Outer Hebrides in December 1967. In this view, looking northeast over the Canadian Atlantic seaboard, Cape Breton Island in Nova Scotia is just visible below the cloud at bottom center.

THE BASIC CONCEPT OF WHAT IS CALLED THE HYDROLOGICAL CYCLE was understood as far back as the sixth century BC. Thales of Miletus (the famous astronomer also interested in weather patterns) was first to realize that water vapor that is evaporated from lakes and seas forms clouds and that it was the same water that fell as rain to replenish rivers, lakes, and seas. Slightly later, in the fifth century BC, another astronomer, Anaxagoras, recognized the process of convection in the atmosphere, and noted that temperatures decreased with height and that the tops of some clouds consisted of ice particles.

These ideas were fundamentally correct, although they were not generally accepted for hundreds of years. We now know that there are two different mechanisms leading to the formation of rain. Cloud droplets are extremely small (0.001–0.05 mm in diameter), so they easily remain suspended in the air and growth by collision is very slow. When there is strong convection inside deep clouds, however, collisions may eventually produce raindrops, which are generally about 0.5–2.5 mm in diameter. This type of rain is common in the tropics and often occurs in temperate regions in summertime when cumulus clouds may reach high into the sky.

At all latitudes, however, there is another process that occurs when clouds grow high enough for the

2 | SHOWERS WITH SUNNY INTERVALS

temperature to drop below 32°F/0°C. The water droplets freeze into ice crystals in the process known technically as "glaciation." This is marked by changes in the tops of the clouds, which become fibrous in appearance. The crystals immediately begin to fall under the influence of gravity and collide with other ice crystals or unfrozen water droplets, growing rapidly as they do so. Generally the crystals melt in the lower layers, becoming large raindrops. If the surface temperature is low, there may be falls of ice crystals or snow.

Most stratiform clouds, apart from nimbostratus, do not normally produce a great deal of precipitation, but they may give rise to drizzle or, if temperatures are low enough, a fall of tiny ice crystals from a higher layer that may "seed" a lower cloud layer, giving rise to bigger raindrops than might otherwise be expected.

Both condensation and freezing require specific nuclei: tiny solid particles that initiate the processes. Condensation nuclei are present everywhere, but suitable freezing nuclei are often absent, when cloud droplets continue to exist as liquid water at temperatures well below freezing—even down to -40°F/C. Such droplets are found in many clouds, especially at high altitudes, as well as in certain fogs, and are said to be supercooled. They freeze immediately when they come in contact with a suitable nucleus or other object (such as an aircraft).

A HEAVY SHOWER FROM A CUMULONIMBUS CLOUD. To the general public, "showers" are gentle, short-lived falls of rain. To meteorologists, however, showers are falls of rain, snow, or hail (often very heavy) associated with cumulonimbus clouds. The gray color here shows that this heavy shower consists of rain, rather than hail or snow.

Hail and rain over Black Rock Desert, Nevada

VIGOROUS UPDRAFTS WITHIN A CUMULONIMBUS CLOUD are required to form hail. Ice crystals that form in the top of a cloud fall into the updrafts and return to higher layers of the cloud. This may happen several times. They gradually grow as they pass through a region where water droplets are supercooled (i.e., exist well below the normal freezing point). Such droplets freeze immediately on contact, adding to the size of the hail. Eventually the particles become too heavy and fall from the cloud. Here, a white hail shaft is surrounded by gray shafts of rain.

Heavy rain over Scotland

TO METEOROLOGISTS, A SHOWER IS PRECIPITATION from an individual cumulonimbus cloud or cluster of clouds, as distinct from the more-or-less continuous rainfall from a frontal cloud. Generally showers cover a restricted area, often no more than 3 miles/5 km across, but multiple-cell storms and supercells cover far greater areas. Rain or hail may be very heavy, especially if (as here over the Sound of Mull) the cloud is moving slowly with the wind. (Note the long, straight, fallstreaks, which only curve close to the ground, where friction has reduced wind speed.)

Previous pages | Heavy rain shower over southern France

ALTHOUGH SHOWERS FROM ISOLATED CUMULONIMBUS CLOUDS affect only a small area at any one time, the amount of precipitation may vary widely. In mature cumulonimbus with strong updrafts, the raindrops tend to be supported by the flow of air, and the amount of rain reaching the ground may be relatively small. As the cloud matures, however, the convection eventually ceases and the updrafts decay. The rush of falling raindrops often produces a powerful downdraft, so a short, intense pulse of rain is accompanied by a powerful gust of wind that spreads out ahead of the cloud.

Formation of dew

WHEN THE SKY IS CLEAR AFTER SUNSET and there is little wind, the ground rapidly starts to radiate heat away to space. When the temperature close to the surface reaches the condensation point, water droplets (dew) are deposited on to blades of grass and other surfaces. If the temperature drops even further, below freezing, water vapor in the air is deposited as ice, giving rise to a coating of hoarfrost.

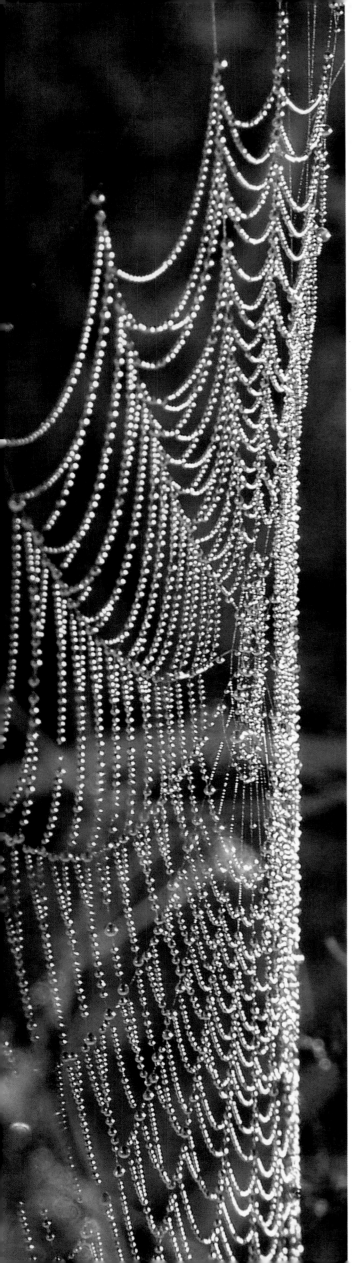

Dewbow

DEW COMMONLY COLLECTS on spiders' webs and, just as with much larger raindrops, the droplets disperse sunlight into a spectrum, similar to a rainbow. This web is vertical, but large dewbows are most frequently seen on extensive areas of grass, covered in a multitude of horizontal webs. Somewhat similar colored bows may sometimes be seen on tiny water droplets lying on the surface film of ponds and puddles.

Guttation drops on a blade of grass

WATER IS TRANSPORTED WITHIN PLANTS from the roots to the leaves. Frequently, when the temperature drops at night, and particularly when the ground remains warm, excess water is carried to the tips of the leaves where it is unable to evaporate into the air and forms large guttation drops—which are often mistaken for dew—as seen here on a blade of grass.

55

Rime "feathers"

RIME ICE AT THE SAENTIS METEOROLOGICAL STATION in the Swiss Alps. Supercooled water droplets—i.e., water that is below the normal freezing point—frequently occurs within the atmosphere, especially when there is a lack of suitable freezing nuclei on which they can form as ice. Such supercooled water droplets freeze as soon as they come into contact with a cold surface, giving rise to a deposit of rime. The rime forms on the windward side, so the "feathers" of rime point into the wind. At high elevations, extremely large deposits several meters across often form on exposed surfaces.

Rime ice "needles"

NEEDLE-SHAPED CRYSTALS of ice, such as those seen here on a wire fence, are characteristic of rime formed from supercooled water droplets. All the needles are on one side of the wire mesh. Hoarfrost deposits, by contrast, form when the temperature of the air drops below 32°F/0°C, and they form on all sides of an object.

Tree covered in rime

SUPERCOOLED WATER DROPLETS can exist at all levels in the atmosphere and, near the surface, often occur as supercooled fog, which freezes instantaneously on contact with trees, shrubs, and other surfaces. The surface temperature when this photograph of a cottonwood tree was taken in the Teton Valley in Idaho was -11.2°F/-24°C.

Frost patterns on window pane

THESE BEAUTIFUL PATTERNS OF ICE arise from the rapid growth of ice crystals from initial, randomly scattered, freezing nuclei. The water vapor in the thin layer of air next to the glass is supercooled and freezes immediately when it comes into contact with any existing ice crystal.

Hoarfrost

HOARFROST FORMS IN A SIMILAR WAY to dew, when the temperature falls on still nights with little wind. Instead of condensing, the water vapor freezes onto exposed surfaces to give a characteristically rough deposit. Because there is little air movement, hoarfrost does not form the "feathers" or "needles" seen with rime, nor the clear layer of ice found with glaze.

59

Extreme desert: Atacama

THE ATACAMA IS AN EXTREME DESERT with minimal rainfall not only over the lower terrain and the coastal ranges but also in the high Andes. Here two snow-covered volcanoes are seen, but the white areas at upper right are not snow but salt pans formed where the meager precipitation has leached minerals from the volcanic soil and then evaporated in the high, dry air. The area is almost completely devoid of vegetation, although some scant traces may be seen along some of the streams, where they appear as red tints (caused by use of a satellite image taken in the infrared region).

Forest fires in North America

EVEN RELATIVELY MOIST CLIMATES, such as that of California and Oregon (shown here) may suffer from occasional severe droughts that can provide conditions suitable for highly destructive forest fires. Here, on July 29, 2002, winds from the interior of the continent have desiccated vegetation to such an extent that it was readily set on fire. The smoke plume (yellow/brown in color) has been swept out over the Pacific. The white lines above the smoke plume are ship tracks, where the exhaust fumes from ships have provided condensation nuclei and created thicker lines of cloud. To the south, the circular pattern of clouds marks the position of tropical storm Elida.

Satellite images of major flooding, China

THESE INFRARED SATELLITE IMAGES OF LAKE DONGTING HU in Hunan Province, China, illustrate the changes during major flooding. In both images, vegetation appears red. In the left-hand image, taken before the floods, the lake appears dark blue. Lighter blue areas are shallow standing water in rice paddies and similar areas. Flood walls are just visible as blue/gray lines. The town of Qionghu appears as a grayish area on the lakeside at bottom left.

IN THE RIGHT-HAND IMAGE, taken at the height of the floods, the main lake, now much larger, appears light blue. The flood walls are now clearer to see and have protected some areas of land. This flood was caused by heavy rainfall in northern China and was exacerbated by erosion caused by deforestation, which had led to silting-up of areas normally available for water overflow. The floods killed over 3,000 people and destroyed some 5 million homes.

FOG AND MIST—WHICH IS JUST A LESS DENSE FORM OF FOG—are the most obvious causes of poor visibility. They arise whenever the air at ground level cools to the condensation point. Three different types of fog are recognized: upslope fog, when humid air is carried uphill by the wind; radiation fog, when the ground radiates heat away to space after nightfall, cooling the layer immediately above it; and advection fog, when humid air is carried across a cold surface, such as the sea or snow- or ice-covered land. This last term is also frequently used for fog that has formed elsewhere and then drifted over a neighboring area. This frequently happens with fog that forms over the sea and then flows inland.

Various areas of the oceans, where a warm airstream flows over cold water, are particularly prone to fog. The Grand Banks in the North Atlantic are notorious, for example, because here warm air from the tropics encounters the cold Labrador Current. Frequent and extensive fogs occur where there are cold oceanic currents and upwelling of cold bottom water from the lowest layers of the ocean. For physical reasons related to the rotation of the Earth, such upwelling takes place along the western side of continents. There are four particular areas where this takes place: two in the Pacific and two in the Atlantic. Off the coast of South America, the Humboldt (or Peru) Current brings frigid water from the Antarctic, and, in the northern Pacific, the California Current carries cold water south along the coast of California. In the South Atlantic, the Benguela Current lies off the coast of Namibia, and, north of the Equator, the Canary Current (as the name implies) carries cold water past the Canary Islands and the northwestern coast of Africa. All these regions are noted for their large fish stocks, fed by the nutrients brought to the surface by the upwelling.

Visibility may also be affected when tiny dry particles are suspended in the air. This arises naturally where winds raise fine dust from the surface, sometimes in dramatic sand- or dust storms, and also as a result of human activity, often caused by deforestation or (in desert areas) by the destruction of "desert pavement," the thin, fragile layer that forms over long periods and binds the finer particles into a more resistant layer. Smoke from deliberate or accidental fires also contributes to haze over wide areas of the world.

Air containing a large proportion of smoke and other pollutants is known as smog. Photochemical smog is especially hazardous and occurs when photochemical reactions, generally powered by sunlight, convert pollutants, particularly vehicle emissions, into highly irritant substances, such as ozone. An inversion can act as a "lid" trapping the smog layer and preventing its dispersal, as happens with the notorious smogs over Los Angeles, for example. Occasionally, photochemical reactions acting on volcanic emissions also create volcanic smog ("vog").

3 | LIMITED VISIBILITY

Previous pages | Atmospheric haze, northern China

THE EXTENSIVE ATMOSPHERIC HAZES that occur over China primarily consist of solid particles derived from smoke and dust. Deforestation of vast areas of loess—an extremely fine-grained soil—farther west has not only caused the whole region to be frequently blanketed by dense haze but has also served as a source of the massive load of sediment transported by the Huang Ho River. This has given rise to the river's alternative name, the Yellow River. In this image, the bay at center left is Bo Hai, and North Korea is at upper left.

Sea mist

SEA MIST AND FOG arise when warm air flows across a cold sea. These conditions often produce extensive areas of fog over the sea, and this may be carried on by the wind to neighboring coasts, giving rise to advection fog. Frequently such fog and mist is confined to the coast because warming over the land causes it to dissipate, giving clear conditions inland. Sea fog or mist tends to persist over the sea, but it clears during the daytime over land and returns at nightfall. Often such fogs have local names as here, at Whitby in North Yorkshire, England, where fog from the cold North Sea is known as "sea fret" or "haar."

Fog over the North Sea

THIS FALSE-COLOR SATELLITE IMAGE of northwestern Europe shows a large bank of fog (pale yellow) over the North Sea. The fog has penetrated inland at various locations along the east coast of Britain. Such conditions arise quite frequently in summer, when there is an easterly flow of warm air from continental Europe that crosses the relatively cold North Sea. High cirrus clouds appear blue, and the snow-covered areas of the Alps, Norway, and Sweden are clearly visible.

Coastal mist over forested dunes at dawn

ALTHOUGH FOG AND MIST ARE IDENTICAL IN NATURE—both may be described as low cloud—
technically, visibility in mist is greater than 0.6 mile/1 km. When visibility is less, it is described as
fog. In many areas of the world, advection fog or mist forms a significant (or only) source of water
for desert areas, such as the Atacama Desert in Chile and the Namib Desert in southern Africa.
This photograph was taken in the Wilderness National Park, South Africa. In southern Africa, the
darkling beetle survives by standing on its head at night, so that dewdrops forming on its body
run down to its mouth, providing it with essential moisture.

Arctic sea smoke

ONE TYPE OF FOG OCCURS when extremely cold air moves over relatively warm water. There is rapid evaporation from the water, causing the overlying layer of air to become saturated, producing tendrils of fog that rise from the water. Arctic sea smoke is also known as steam fog or frost fog. It is not confined to polar regions, although this image was obtained near the Ronne Ice Shelf in Antarctica.

Atmospheric pollution over western Europe

GENERAL ANTICYCLONIC (HIGH-PRESSURE) CONDITIONS over Europe, with clear skies and an
inversion (see page 33), have trapped a pollution haze in the lowermost layer. It is clearly seen
over the Czech Republic, Germany, France, and Great Britain, and it is obvious that it has been
spread toward the north and west by the southeasterly winds.

Smoke plumes from land clearance

OBTAINED FROM ONE OF THE SPACE SHUTTLE MISSIONS, this image shows smoke plumes from extensive fires, which were set to clear marshland in the Tengah region of southern Kalimentan (Indonesian Borneo) to make way for agricultural land, following a surge in population of emigrants from Java. Such land rapidly becomes infertile, which is having devastating effects on all tropical rainforests. The smoke contributes to extensive hazes that have blanketed most of southeastern Asia in recent years.

Saharan sandstorm

A SANDSTORM (WHITE) IS IN PROGRESS over the Djourab region of the Sahara Desert, Chad, at upper center. The prevailing winds in this region are the Northeast Trades, which blow toward the southwest and have created large linear features on the plains. This view is looking southwest, where the winds are channeled between the dark mountainous regions of the Tibesti Mountains (center right) and the Ennedi Massif (center left). The Tibesti Mountains are the highest point in the Sahara Desert, and their highest peak (Emi Koussi) reaches 11,204 ft/3,415 m.

Sandstorm in Qatar

THIS STRIKING IMAGE OVER QATAR SHOWS a sandstorm sweeping south (to the right) toward Saudi Arabia and the United Arab Emirates. The sandstorm was triggered by strong winds around a deep depression over southwestern Asia. There is a particularly well-marked boundary between the clear, undisturbed desert to the right and the turbulent clouds of sand and dust to the left.

4 | A COLD SNAP

climate and weather of surrounding regions. The vast ice sheets and glaciers that covered large areas of the globe during the last ice age are largely responsible for the landscape we see today, particularly in Canada, northern Europe, and Asia. During the current period of warming, some glaciers have retreated dramatically by tens of kilometers when compared with their positions a century or more ago. Some, however, are advancing, because the warming trend has produced increased precipitation in their accumulation zones.

Ice shelves and pack ice float, of course, so the increase or decrease in their total area has no direct effect upon sea level. The area covered does, however, have a strong influence on ocean currents and on the weather of surrounding regions. An increase in the amount of freshwater released by melting of parts of the ice caps covering Antarctica and Greenland (especially the latter) could have a major impact on the oceanic circulation and could lead to a significant deterioration of the climate of western Europe in particular.

On a more immediate level, heavy snowfall and blizzards can cause major disruption to transportation and everyday life. Ice storms (more properly described as the widespread incidence of freezing rain that covers any exposed surfaces with large accumulations of glaze) may be highly destructive. The ice storm in Canada and New England on January 5–9, 1998, is a relatively recent example of the ensuing widespread disruption and damage.

THIS UNUSUAL ICEBERG is locked in frozen sea ice but at an earlier stage must have been floating freely in the water. An arch was created in the iceberg in the form of a giant hole, the base of which is hidden below the sea surface, which has subsequently frozen. As always, the largest portion of the iceberg is below the surface.

Hailstorm, Gillette, Wyoming

HAIL CONSISTS OF SOLID BALLS or pellets of ice. It forms in cumulonimbus clouds with vigorous convection and the consequent strong updrafts. Beginning as raindrops or ice crystals, particles are swept up into the colder regions of the cloud by powerful updrafts, where they collide with supercooled water droplets or other particles. They then begin to fall but are generally captured once more by the updrafts to complete another cycle. Eventually they become so large that they cannot be supported by the rising air and they fall out of the base of the cloud, sometimes even being "thrown" downward by strong downdrafts.

Hailstone aggregate

AS IF INDIVIDUAL HAILSTONES were not large enough, they often clump together to form hailstone aggregates, which are even more damaging and dangerous. The largest such aggregates known fell at Hyderabad in India in 1939, with a weight of 8 lb/3.7 kg, and at Yūwu in China in 1902, where the weight was as much as 9 lb/4 kg.

Hailstones

SOME OF THESE INDIVIDUAL hailstones show indications of their internal structure, which consists of concentric layers of opaque and clear ice. When the growing stones pass through cloud layers in which there are relatively warm liquid water droplets, the water spreads out to form a clear layer of ice. In other layers where the droplets are supercooled, they freeze instantly on contact and air is trapped between the individual tiny droplets, giving rise to an opaque layer. Hailstones may reach 4 in/10 cm across in severe storms. The record weight is 2.2 lb/1 kg, for a single stone about the size of a cantaloupe melon, which fell at Gopalganj in Bangladesh in 1986.

Ice crystals under polarized light

THIS MICROGRAPH OF A THIN SECTION of ice reveals a whole range of colors. These patterns appear because different wavelengths of plane-polarized light are rotated by different amounts. The range of colors indicates that there is considerable strain within the ice. This tension arises because the outer layers froze before the inner layers, introducing considerable stress into the crystal structure. It is this internal stress that sometimes leads to hailstones "exploding" when they hit the ground.

Impending blizzard

THIS CAMP IN GREENLAND is just about to be hit by a heavy snowstorm and high winds. Although the location of the Sun may still be seen, it will soon become completely invisible. Such conditions are popularly called a blizzard, but strictly speaking, the term should be reserved for high winds (force 7 or more on the Beaufort scale) that are laden with blowing snow raised from the surface, rather than falling snow. In either case, however, visibility is usually extremely low, rendering any travel almost impossible.

Avalanche, west shoulder of Mount Everest

AVALANCHES MOST COMMONLY OCCUR because of the development of a weak layer of snow, known as depth hoar, at some distance below the snow surface. Any slight shock (such as a rockfall, an earthquake, thunder, or human activity) may be sufficient to set the uppermost layer into motion. This is a powder avalanche, caused by the collapse under its own weight of a deep layer of soft, poorly consolidated snow. Two other, distinct forms of avalanche are the slab avalanche, when a large, relatively coherent layer becomes detached from the underlying layer, and a wet avalanche, where the collapse is initiated by thaw, and which has a high density and may be extremely large.

Glaze or "black ice"

WHEN SURFACE TEMPERATURES are below freezing, a depression may introduce relatively warm air at a higher level. Rain, falling from the warm layer, freezes on contact with objects on the ground and covers them in a layer of clear ice, known as glaze, and more popularly called "black ice." Here, rain falling on a road sign has created a perfect octagon of ice, which has slipped down from its original position.

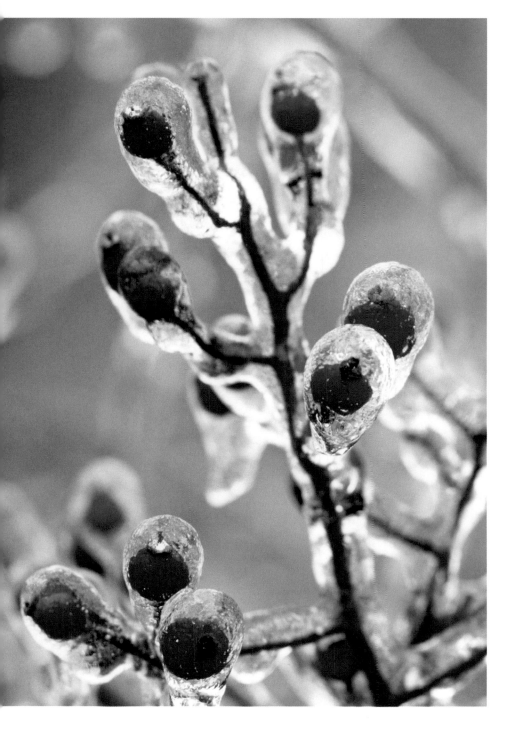

Rose hips encased in glaze

A LAYER OF GLAZE IS COATING these rose hips after a winter storm that swept across Michigan. It is easy to see how the ice forms a relatively even coating all around the hips and stems of the rose.

97

Hubbard Glacier, Alaska

THIS IS A LARGE GLACIER, and the cruise ship below the center of the image gives some sense of scale. The heavily crevassed and weathered upper surface of the glacier is clearly seen. The crevasses and the effect of the tides, which put considerable strain on the ice, cause large sections to break away to form icebergs.

Ogives on the Gilkey Glacier, Alaska

ANNUAL GROWTH CURVES, OR OGIVES, on a glacier consist of alternating light and dark bands. Raised bands occur where summer warmth has melted some of the surface ice. When the surface refreezes in winter, it is darker because of accumulated dirt and debris. Winter growth produces just clean white ice. The darker ice absorbs more sunlight and partially melts, creating shallow troughs in the surface. The ogives are convex downstream, because the sides of the glacier move more slowly than the ice in the center. The dark streaks parallel to the glacier's direction of movement consist of debris from the rocks at the valley sides.

Crevasses on an ice shelf, Antarctica

THESE DEVELOP WHERE THERE ARE CHANGES in the rate at which the ice is flowing, as a result of friction with the sides of a valley, changes in the width of the valley, or where there are significant changes in the level of bedrock beneath the ice. When the changes are especially pronounced, particularly because of a major, abrupt change in the level of the bedrock, the ice may be so fractured that the area is known as an icefall. On ice shelves, crevasses generally arise because of the continuous flexure of the floating ice due to tides and oceanic swells. This pavementlike pattern is found on the ice shelf in George VI Sound.

Sastrugi on ice cap, Heer Land, Spitzbergen

SASTRUGI (LEFT) ARE FORMED WHEN hard particles of ice scour the surface, driven by high winds. They carve the surface into a wide range of forms, including grooves that may be as much as 6.5 ft/2 m deep. In this respect they resemble yardangs (see page 241), the grooves cut into weak sedimentary layers of soil by the action of the wind. Sastrugi usually form in dry snow when temperatures are below 14°F/-10°C.

Penitents, Alaska

PENITENTS (also frequently known by the Spanish term *nieve penitente*) arise when the surface of a glacier—which may be snow, firn, or solid ice—becomes heavily fractured. Individual pillars waste away to give rise to tapering spires of ice such as these on the Hole in the Wall Glacier, Alaska.

101

Meltwater cave, Worthington Glacier, Alaska

WATER COMMONLY PENETRATES CREVASSES and accumulates at the base of a glacier, escaping through a cave in the snout. Such streams may be very strong and carry large quantities of debris. When glaciers retreat, the retreating meltwater caves leave behind snaking lines of rubble, known as eskers. Eskers created by the retreating ice caps of the last ice age, and now colonized by vegetation, are often the most prominent features in otherwise flat, eroded landscapes, such as those found in Canada and Finland.

Stranded iceberg, northwestern Greenland

THIS ICEBERG HAS RUN AGROUND in shallow water, and the portion normally underwater has been
exposed by low tide, together with the flutes created by its slow melting in seawater. It also shows
that, although most of the mass of an iceberg may be under the surface, it may not be any wider
underwater than it is above the surface.

The Tyrolerfjord, Greenland

THIS FJORD IN NORTHEAST GREENLAND consists of a flooded valley cut by an Ice-Age glacier. It illustrates the characteristic U-shaped profile of all glacial valleys, which differs from the distinct V-shape found in valleys cut by rivers. The majority of fjords also have a definite step, or underwater bar, at the seaward end, where the ice causing the erosion lifted away from the bedrock upon reaching the sea. The photograph was taken from the summit of Ehrenberg Fjeld in the North East Greenland National Park.

The West Fjords (Vestfirdir) Peninsula, Iceland

THIS IS AN EXTREME EXAMPLE of the type of jagged and indented coastline created by glacial action and the subsequent flooding following sea-level rise at the end of the Ice Age. A remnant of the former deep snow and ice cover is still found on the mountain peaks, but the subsidiary valleys that once held permanent glaciers are now clearly ice-free. The large Arnarfjordur inlet is at lower left.

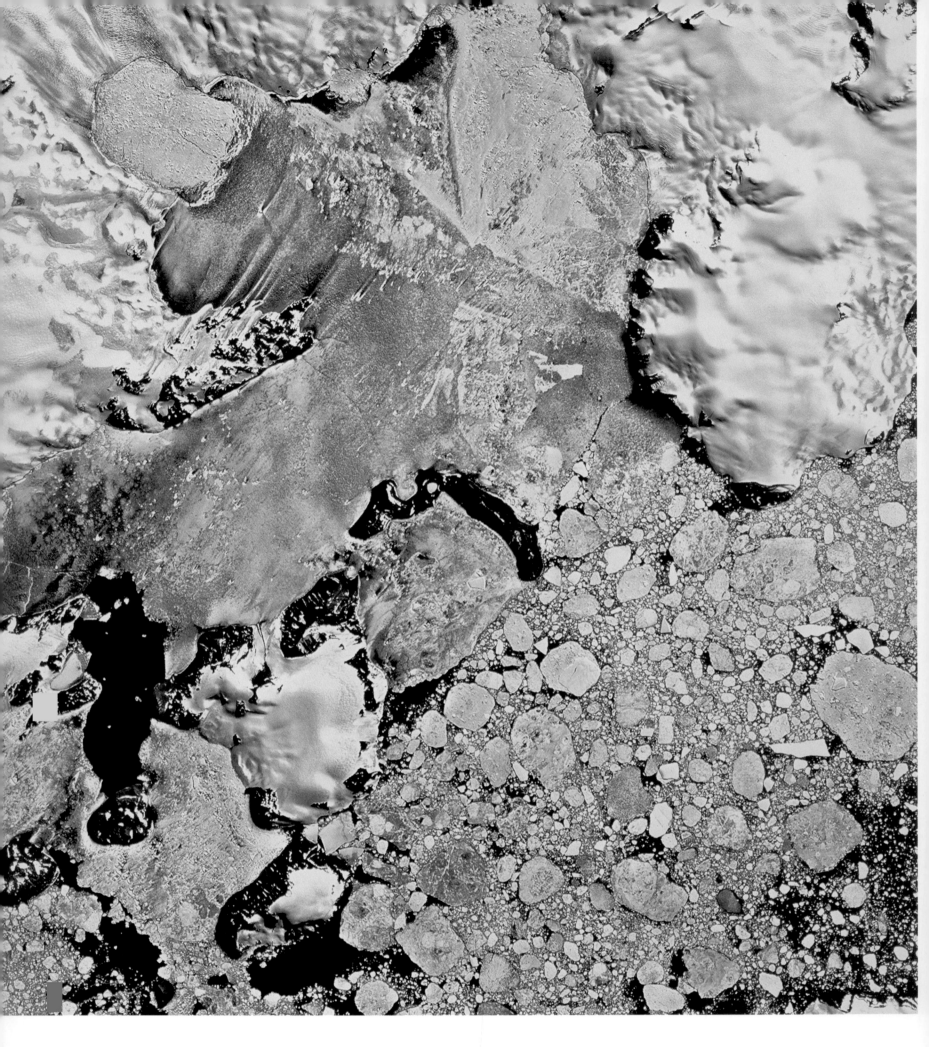

Islands and sea ice, Trinity Peninsula, Antarctica

THE SNOW-COVERED MOUNTAINS of Trinity Peninsula are top left, Vega Island to the right, and a cluster of smaller islets (some appearing dark with white flecks of snow) are at bottom left. Much of the image is covered by melting shelf ice (bluish tints) which is breaking up into pack ice at the bottom of the image, taken in February, in the southern summer.

Sea ice, Labeuf Fjord, Antarctica

THIS SUMMER IMAGE ILLUSTRATES the complex patterns that occur when sea ice starts to break into ice floes. Initially all the ice was fast ice, that is, ice either frozen to the shore or the seafloor, or else grounded in shallow water. The speckled appearance of the ice is caused by innumerable pools of meltwater on the surface, which appear black. The ice itself is becoming waterlogged, and cracks (white) are beginning to appear, which will eventually widen to create numerous individual floes. Large areas may suddenly break free from the shore or the floor of the fjord and be carried out to sea by the wind.

EXTREME WEATHER EVENTS CONSTANTLY HIT THE HEADLINES, often prompting the question of whether they are caused by global warming. Although practically all meteorologists and climatologists are certain that global warming is taking place, such is the nature of science that no one can say positively that extreme events have one specific cause. It is often found that a similar natural event occurred in the past. It is also true that the longer you observe any phenomenon, the more likely you are to record extreme events. Weather records obtained with accurate instrumentation date back only about 150 years, so it is not surprising that occasionally events occur that break all previous records.

Modern communications are essentially instantaneous, and this adds to the perception that more extreme events are occurring than happened previously, when news traveled more slowly around the world. But it is precisely this speed of communication and the worldwide distribution of observers and observing stations that are now giving meteorologists the means to make reasonably accurate predictions of extreme weather events. Even though the exact mechanisms responsible for the production of lightning and tornadoes—to give but two examples—are still uncertain, enough is known about atmospheric physics for forecasters to issue "watches" when severe thunderstorms or tornadoes are possible, and specific "warnings" when they are imminent.

Similarly, great advances have been made in forecasting the tracks of tropical cyclones (also known as hurricanes, typhoons, or cyclones), which are so large and cause damage and destruction over such large areas that the evacuations of tens or hundreds of thousands of people may be required. Any improvement in the accuracy of such forecasts will not only save lives but also minimize the disruption that is caused. Tropical cyclones are accompanied by storm surges, but these may also be created by less extreme weather, where the winds accompanying deep depressions may combine with high tides to overwhelm sea defenses. Again, it has now become commonplace for such storm surges to be predicted with considerable accuracy.

Major floods that occur through the melting of winter snowpack or excessive winter rainfall are generally predictable some time in advance, although their precise extent and timing depend so greatly upon local factors that the exact details remain uncertain. The prediction of flash floods, which may cause extensive damage and loss of life within a restricted area, is far more difficult to achieve. Disastrous floods such as the Big Thompson Canyon flood in Colorado on July 31, 1976, in which 237 people lost their lives, or the Lynmouth, Devon, England, flood of August 15, 1952, when over thirty people died, were so localized that no specific warning could be given (nor could one be given nowadays), other than that conditions were such that torrential rain and possible flooding could occur. Now, however, the use of weather radar provides a method of determining where extreme events are occurring in real time, thus giving the emergency services valuable additional information.

5 | WEATHER WARNING

Previous pages | A supercell thunderstorm

WITHIN A SUPERCELL THUNDERSTORM, a single massive rotating updraft (known as a mesocyclone) lies in the heart of the cloud. Such supercells persist for many hours and produce violent weather, such as this thunderstorm over Kansas farmland, including extremely heavy rain and hail, and frequent lightning. They also often spawn violent tornadoes.

Lightning

TWO OF THE HAZARDS OF SUPERCELL STORMS. Although supercell storms produce numerous lightning strokes and occasional tornadoes, it is rare for both to be captured in a single image, as happened here in Florida. Both are clearly visible, because there is no rain or hail falling to obscure the view. Tornadoes often descend from an area of cloud known as the "rain-free base."

Global distribution of lightning

THIS DIAGRAM OF THE ANNUAL FREQUENCY of lightning discharges per square kilometer ranges from red/orange (high), through yellow, green, blue, purple, and gray, to white (low). For many years, Indonesia was believed to have the highest concentration of lightning strikes, but satellite observations revealed that the region around Kampala in Uganda held the record. Other "hot spots" occur over Florida, the northwestern Himalayas, and Colombia.

Cloud-to-cloud lightning

Cloud-to-ground lightning

A LATE-EVENING VIEW of a major thunderstorm over Phoenix, Arizona. Lightning discharges not only occur from cloud to ground but also within a cloud, between different clouds, and between cloud and clear air. The mechanism by which positive and negative charges become separated is still imperfectly understood, as are the fine details of precisely how discharges take place. Lightning strokes, such as the one shown here, which runs across the base of the cumulonimbus anvil, are known as "anvil crawlers" to storm chasers.

LIGHTNING FLASHES NEAR STERLING, COLORADO. The initial channel is created by what is known as a stepped leader, and in most cases this brings negative charge from the cloud down to the ground. The main positive stroke then flows upward into the cloud. Multiple strokes may follow the same initial channel, consisting of dart leaders from the cloud and return strokes from the ground. In a small percentage of cases (about 5 percent), positive charge descends from the cloud, producing a negative return stroke that carries a far greater current.

Lightning strikes near Kourou, French Guiana

TWO METEOROLOGISTS STUDY the distribution of lightning near the Ariane rocket launch site (marked "ELA") over a period of one month. Knowledge of the likelihood of severe weather and lightning in particular is essential in planning satellite launches. The jagged white line running diagonally across the screen indicates the coastline, and it is noticeable how almost all the lightning strikes occur over land.

Artificially triggered lightning

AT SOME RESEARCH STATIONS, particularly in France and the United States, lightning is artificially triggered. A rocket is launched into a thunderstorm trailing a fine copper wire. This provides an easy path for the lightning discharge to reach the ground. Such research allows scientists to measure the current, voltage, temperature, and other parameters of lightning strokes.

Steam or vapor spout

AT LOCATIONS (SUCH AS HAWAII) where molten lava reaches the sea, it creates large amounts of steam, which may form a localized cloud in the form of a steam or vapor spout. Occasionally, conditions may be right, with strong updrafts and humid air, for the formation of a rotating funnel, within which the reduced pressure causes the vapor to condense. Such steam spouts are normally very short-lived, lasting no more than a few minutes.

Gustnado, New Mexico

A GUSTNADO is a relatively weak whirl that develops from the outflow that occurs along the gust front from a thunderstorm, line squall, or other storm system. The mechanism by which it forms is similar to that of a dust devil or a landspout, rather than a true tornado, but it may still be strong enough to cause some damage.

Tornado formation and development

THESE IMAGES ARE TWO VIEWS OF A TORNADO moving across Kansas farmland. Major tornadoes form within an area beneath supercell storms known as the rain-free base, which is the site of the most powerful updraft, or from a wall cloud, which is an even lower region that often exhibits strong rotation. Indications of the wall cloud may be seen in both photographs.

THE FIRST PHOTOGRAPH (far left) shows the low pressure in the center of the tornado causing water vapor to condense, resulting in a funnel cloud hanging from the cloudbase that is clearly visible. This condensation funnel is often obscured by the material raised from the surface. In weak tornadoes, such as those that form beneath nonsupercell storms, the presence of such a debris cloud may be the only evidence for a tornado, with no funnel cloud present.

127

ONCE CONTACT HAS BEEN MADE with the ground, the diameter of the tornado may expand greatly—as seen in the photograph near left—but the footprint may vary considerably during a tornado's lifetime, both because of the changing altitude of the ground and because the strength of the vortex fluctuates.

Hurricane Mitch

THIS COLORED THREE-DIMENSIONAL computer image of Central America and the highly destructive Hurricane Mitch is based on satellite data for October 26, 1998, three days before the hurricane made landfall on Honduras. Honduras itself is almost completely hidden by the shield of cirrus clouds above the hurricane. It produced winds of over 199 mph/320 kph and torrential rain that caused flooding and landslides. Over 11,000 people were killed, and damage was estimated at $5 billion.

Multiple-vortex tornado, South Dakota

THIS MULTIPLE-VORTEX TORNADO was one in the major outbreak in South Dakota on June 24, 2003. In major tornadoes, particularly those with an extremely large footprint, subsidiary whirls, known as suction vortices, may form around the base to create a multiple-vortex tornado. Tornadoes tend to leave evidence of their passage in the form of swirls on the ground, and the secondary vortices may sometimes be revealed by the small, tighter marks that they produce.

Tropical depression from space

TROPICAL CYCLONES (also known as hurricanes, typhoons, or cyclones, depending on where in the world they occur) develop in a series of stages. Beginning initially as tropical disturbances— areas of low pressure and organized convection—they may develop into tropical depressions, like the one shown here off California. These consist of a series of cloud bands surrounding a low-pressure center. When the circulation becomes even stronger and more organized, the system may be classified as a tropical storm, the final stage before the development of a true tropical cyclone.

Tropical storm Iniki

THIS STORM ORIGINATED AS HURRICANE INIKI, which caused serious damage on the Hawaiian island of Kauai in September 1992. Winds in the hurricane then exceeded 149 mph/240 kph and generated waves of over 16.4 ft/5 m in height. When the hurricane passed over Hawaii and into the cooler waters of the North Pacific, it began to weaken and was downgraded to a tropical storm, seen here from space. This weakening is shown by the lack of strong convection and thunderstorm activity near the center of the cloud spiral and by the breaks in the cloud cover farther out.

Tropical cyclones and tropical storm

THESE TROPICAL CYCLONES and a tropical storm can be seen over the Indian Ocean. The coast of Madagascar appears at far left. The systems visible here, in February 2003, are (from left to right): tropical cyclones Gerry and Hape, tropical storm 18S, and tropical cyclone Fiona. Tropical cyclones only form over tropical oceans where the sea surface temperature exceeds 80.6°F/27°C. Most systems then track westward until they either cross onto land or move over cooler waters. In both cases, their energy source is cut off, and they gradually decay.

Typhoon Violet over Japan

SOUTH KOREA APPEARS AT TOP LEFT in this picture: a representation of wind speeds and directions in typhoon Violet as calculated from satellite data. Wind direction is indicated by the arrows and wind speed by the color, from blue (6.6 ft/s/2 m/s) to red (more than 65.6 ft/s/20 m/s). The winds spiral around a central clear, calm eye where the air pressure is extremely low. Typhoon Violet moved over Japan on September 21–22, 1996, causing a number of deaths and considerable damage.

Motion of Hurricane Andrew

A COMPOSITE IMAGE SHOWS the motion of Hurricane Andrew over three days (August 23–25, 1992). Andrew, a category 4 event, had winds of 143 mph/230 kph and a central pressure of 922 millibars. Only Camille in 1969 and an unnamed hurricane in 1935 had lower central pressures (909 and 892 millibars, respectively). Andrew was also the costliest hurricane up to then, causing damage amounting to over $26 billion, more than three times that caused by Hugo in 1989.

Cloud shield and eye of Typhoon Fefa

THIS VIEW FROM THE SPACE SHUTTLE was taken in August 1991 when the tropical cyclone was 621 miles/1,000 km east of Taiwan. The eye, where air descends in the center of all tropical cyclones, is particularly well defined, with so little cloud that the sea surface is actually visible. The wall of the eye extends from about 0.6 mile/1 km above the sea to well over 8.7 miles/14 km.

Typhoons near Japan

SYNOPTIC VIEW OF WINDS over the northern Pacific Ocean showing two typhoons near Japan. Wind speed is shown by colors (blue lowest, yellow highest), and wind direction is shown by the white streamlines. Land areas are gray. Typhoon Violet is the system just below Japan (far left) and Typhoon Tom is the larger system at upper center. Because of the size of the Pacific Ocean and the long paths possible over warm waters, typhoons may become gigantic and more powerful than systems over other oceans. Such "supertyphoons" include Typhoon Tip in 1979. It reached a diameter of 1,044 miles/1,680 km and had the lowest central pressure ever recorded: 870 millibars on October 12, when west of Guam.

WEATHER

Radar tracking of hurricanes

THIS PHOTOGRAPH SHOWS a radar screen on board a Hercules aircraft "Hurricane Hunter" about to penetrate the eyewall of Hurricane Floyd in September 1999. The clear eye of the hurricane is readily visible on the screen. The radar pulses are returned by raindrops and hailstones, so the image maps the regions where these are most intense. Some of the rain bands and the clear intervals between them are also visible on the screen.

Storm surge, Hurricane Isabel

THE DESTRUCTIVE POWER of a storm surge is shown as it destroys North Carolina State Highway 12. The highway was swept away, transforming Cape Hatteras Village into an island. Storm surges are immensely destructive: huge mounds of water are raised when tropical cyclones hit the coastline, particularly when the latter is gently shelving. The greatest height ever recorded for a surge was 40 ft/12.2 m at Hatia Island, Bangladesh, on November 12, 1970.

ALTHOUGH ALMOST EVERYONE HAS SEEN A RAINBOW, few people realize that this is just one of many optical phenomena to be seen in the atmosphere. Indeed, not all rainbows are the same: they actually display different forms and subsidiary features that normally go unnoticed. Fogbows and dewbows are two other related phenomena. Halos, created by sunlight or moonlight refracted or reflected by ice crystals, are often missed even though it is estimated that at middle latitudes a halo is visible, on average, every three days. Although there are just a few common halo phenomena, there are also dozens of rarer forms. Many are seen only in polar regions, particularly in the Antarctic, where fine ice crystals ("diamond dust") suspended in the air provide ideal conditions for the creation of halo effects.

Some phenomena, such as coronas and iridescence, are commonly unseen because they occur close to the sun and are lost in its glare but become visible if the sun is hidden. Coronas, in particular, are more frequently seen around the moon, and in fact there are lunar forms of most optical phenomena, with lunar rainbows and halos being quite common.

Certain optical effects have become more common with the advent of widespread air travel. Glories, for example—colored rings seen on clouds or mist around the shadow of an observer's head—were once seen only by mountaineers but are now frequently visible around the shadow of the aircraft an observer is traveling in.

The same applies to subsuns—elongated brilliant patches of light, caused by ice crystals—which at one time were only regularly seen in the polar regions. (Glories may be more common nowadays, but an exact mathematical explanation of the mechanism creating them is still unknown.)

The blue of the sky arises through scattering of sunlight by air molecules. The yellow and red skies of sunrise and sunset occur when blue light has been scattered aside, leaving just the longer wavelengths. On very rare occasions, when there have been major volcanic eruptions, particles ejected into the upper atmosphere contribute red light, which mixes with the blue to give a vibrant purple hue to the sky.

One great natural spectacle is provided by the aurora, which occurs high in the atmosphere, between altitudes of 62 and 621 miles/100 and 1000 km. Although displays are generally seen at high latitudes, in what are known as the auroral zones, roughly centered on the Earth's magnetic poles, on rare occasions displays extend as far as the magnetic equator. Occasionally, auroras may thus be seen from any place on Earth. Some displays are quiescent, but many exhibit an astounding variety of shapes, colors, and motions. Although auroras may exert a minor, and indirect, influence on weather through their heating of the upper atmosphere, the magnetic storms that accompany major events can damage satellites, disrupt radio communications, and cause electrical surges and blackouts by affecting power lines at ground level.

6 | ATMOSPHERIC OPTICS

THIS HOUSE IN WOOLSTHORPE is the birthplace of Sir Isaac Newton, who first gave an accurate description of how the colors arose in a rainbow. The most common rainbow is the inner (primary) bow, but the fainter outer (secondary) bow, with a reversed color sequence, is not rare. The center of the bows is always the "antisolar point," the point on the sky directly opposite of the Sun, relative to the observer. The primary bow arises through a single reflection from the back of falling raindrops, while the secondary bow undergoes two reflections within the same raindrops. "Alexander's dark band" is an area between the bows where light is directed away from the observer.

Lunar corona seen through thin cirriform clouds

CORONAS OCCUR AROUND BOTH THE SUN AND THE MOON, but because of the glare from the Sun, lunar coronas are usually much easier to see. This image shows the inner aureole, with a brownish ring on the outer edge. Additional colored rings occur farther away from the center, and these are the ones generally seen around the Sun. The rings arise from diffraction of sunlight by particles in the cloud.

148

Lunar corona created by pollen

CLOUD PARTICLES ARE NOT UNIQUE in causing coronas. Here, a beautifully colored corona has been created by an extremely high concentration of pollen grains, released by pine trees in spring in Finland. It s possible, although not confirmed, that the slightly elliptical shape of the corona has been created by the specific shape of the pollen grains. Such coronas are quite common in heavily forested areas, particularly in the northern climatic zone known as the taiga with its vast coniferous forests. Pollen from birch, the hardiest of the deciduous trees, may also create the same effect.

Auroral arc

A QUIET DISPLAY OF THE AURORA BOREALIS. The aurora borealis (northern lights) and the southern counterpart, the aurora australis, are created when highly energetic charged particles, which originally arise in the Sun, cascade into the upper atmosphere, causing the atoms and molecules to become excited and emit light. In this case, the form of aurora is known as a homogeneous arc—quite frequently seen early in a display—although suggestions of a rayed structure are beginning to be evident, a sign that the display is becoming more active.

Auroral corona

THE CHARGED PARTICLES from the Sun penetrate the outer regions of the Earth's magnetic-field region (the magnetosphere) and accumulate in the central plane of a magnetic "tail" that extends out away from the Sun. When the Sun is particularly active, the particles are accelerated up the tail, toward the Earth, and, following the magnetic-field lines, crash into the upper atmosphere in zones around the magnetic poles. An auroral corona appears when the observer is looking directly along the magnetic-field lines.

Red aurora borealis display

THE COLORS SEEN DURING AURORAL displays depend upon the energy of the charged particles that enter the upper atmosphere and the atoms or molecules that are excited. Red displays, such as this one, arise from the excitation of oxygen at high altitudes. A slightly different color is produced when highly energetic particles penetrate to lower altitudes and excite nitrogen molecules. In the past, such displays were often taken to be distant towns or cities on fire.

Auroral rayed band

THIS STRIKING AURORAL RAYED BAND was photographed from Finland. An auroral band appears as a ribbon of light across the sky. This particular band displays a rayed structure, indicating the lines of the magnetic field. The color at the bottom of the display arises from oxygen, which normally appears green rather than the yellowish tint seen here. The violet coloration at the top of the rayed band arises from ionized nitrogen molecules that are not only excited by the charged particles entering the atmosphere but which also lie in sunlight.

"Auroral green" band

DURING AN ACTIVE DISPLAY, the aurora often develops striking bands. These curtainlike folds frequently show rapid motion, reinforcing the illusion of gigantic curtains waving in the sky. Sometimes several, roughly parallel, bands are visible simultaneously. In this display, photographed over Manitoba in Canada, the green coloration is the highly characteristic "auroral green," created by a so-called forbidden oxygen emission, i.e., one that only arises when the density of the gas is extremely low, as it is in the upper atmosphere.

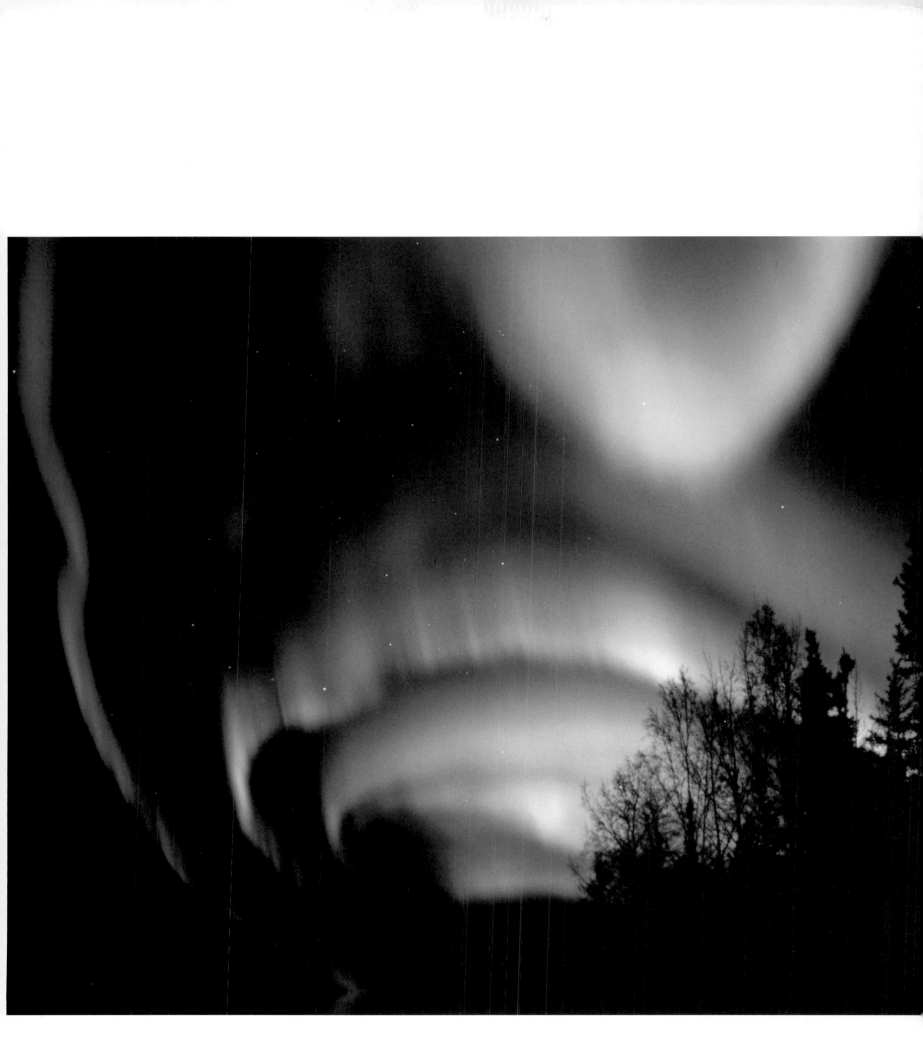

Multiple auroral bands

ALTHOUGH MANY AURORAL DISPLAYS are quiescent, sometimes hardly changing for long periods, some displays are highly active changing shape and position very rapidly. Some, such as this display over Fairbanks, Alaska, become highly complex with multiple bands, where separate sheets of charged particles are entering the upper atmosphere. In general, auroral displays occur at an altitude of between 62.4 and 621.4 miles/100 and 1000 km.

Rayed band

AURORAL DISPLAYS TEND TO START QUIETLY, with irregular patches of light, which may then develop into an evenly glowing homogeneous arc reaching across the sky. Many events do not develop further, but active displays may go on to exhibit a rayed arc, a band (or bands), and rayed bands. Generally the most active displays show rapid motion, often with sudden changes of brightness.

Auroral oval

THIS IMAGE FROM THE POLAR SATELLITE shows an almost complete auroral oval—the region where charged particles are entering the atmosphere at any one time. The aurora is most active in the area that is a reddish brown and weaker (yellow, blue) elsewhere. The greatest activity occurs shortly after midnight, but some displays are so strong that auroral light may be detected by satellites even around midday, when the display is invisible from the ground.

Aurora australis from space

THE AURORA AUSTRALIS, photographed from the space shuttle. During an auroral event, similar displays occur simultaneously at both ends of the magnetic lines of force that link the northern and southern hemispheres. The shuttle's low orbital altitude (less than 310.7 miles/500 km) frequently carries it through the center of auroral displays.

All-sky image: aurora borealis

FISH-EYE LENSES are often used to obtain regular images of the whole sky to monitor auroral activity. Auroral activity commonly occurs in the auroral zone, a region approximately 15–30° from each magnetic pole, where auroras are more frequently seen. During major solar storms, the zone expands and may even reach as far as the magnetic equator.

satellites offer a truly global coverage, day and night, throughout the year.

Imaging the weather: the old...

THIS IS THE VERY FIRST MOSAIC of global weather systems ever obtained, using data returned by the experimental Tiros 9 polar-orbiting satellite, showing the situation on February 13, 1965. Despite its poor resolution by modern standards, it heralded a revolution in the way meteorologists studied weather and climate.

Previous pages | Weather around the world

THIS MOSAIC WAS CREATED from numerous satellite observations showing the weather around the world. The colors and shading approximate those that would be seen by an observer in space. The land is dominated by the green areas of vegetation and the brown tints of the arid lands in the tropics. The clouds were observed over a period of two days.

...and the new

A MOSAIC OF GLOBAL WEATHER systems produced from data returned from five geostationary satellites. With continuous coverage of the whole world, such images are now available every thirty minutes—in this case for 15:00 GMT on January 18, 2005. Because half of the globe is in darkness at any time, observations in the infrared were used as the principal source of data, and the colors were computer-generated to approximate those that would be seen by the human eye.

The Earth's cloud cover in 3-D

THIS COMPOSITE IMAGE OF THE EARTH showing surface temperature (various colors) and three-dimensional cloud cover (white), was created by integrating multiple satellite data, and is the first representation of the Earth's cloud cover in three dimensions. Temperature is color-coded, from red (high) through to dark blue (low). A strong El Niño event in progress in the Pacific Ocean is visible as a red band in equatorial regions. The height of the clouds is greatly exaggerated.

Geostationary satellite image: Africa and Europe

GEOSTATIONARY SATELLITES ARE STATIONED 22,308 miles/35,900 km above the equator, where their orbital period is precisely equal to twenty-four hours, so they remain essentially stationary above one point on the Earth (because the Earth's gravitational field is not uniform, they do drift slowly out of position, and this is corrected by the use of thrusters). This Meteosat image is centered on longitude 0° (the Greenwich Meridian) and latitude 0°. Satellites observe the Earth at various regions of the spectrum, each of which gives a black-and-white image. Channels may be combined to give images with approximately correct colors.

Geostationary satellite image: North and South America

THIS FALSE-COLOR IMAGE WAS OBTAINED from the GOES-E satellite, located over the equator at longitude 75°W. Apart from Meteosat at longitude 0°, three other geostationary satellites provide complete coverage around the world: GOES-W at 135°W, GOMS at 76°E, and GMS at 140°E. Although coverage is excellent in tropical and subtropical regions, the curvature of the Earth introduces limitations at high temperate latitudes and over the poles. Monitoring of these regions is carried out by polar-orbiting satellites at a much lower altitude.

Western Europe from a polar-orbiting satellite

POLAR ORBITERS PASS AT MUCH LOWER ALTITUDES (497–621 miles/800–1000 km) than geostationary satellites and have highly inclined orbits to cover the polar regions. As the Earth rotates beneath the orbit, successive passes cover adjacent (overlapping) swathes of the surface, with two overhead passes of a particular region every day. In this image of western Europe, an anticyclone (a high-pressure area with cloudless skies) lies over the British Isles. The swirl of clouds marks a depression (low-pressure area) over the Atlantic.

Tracking hurricanes: Hurricane Katrina

SATELLITES PROVIDE VITAL INFORMATION in monitoring the position of tropical cyclones (known as hurricanes over the Atlantic). Here, Hurricane Katrina, with its vast swirl of clouds and central eye, is seen on September 28, 2005, just before it hit Lousiana and Mississippi, devastating the Gulf seaboard and inundating New Orleans.

True-color image of Ireland

THIS IMAGE WAS TAKEN BY THE TERRA SATELLITE, which carries instrumentation specifically designed to monitor various factors that may be involved in climate change. Pictured is the rugged coastline of western Ireland, with its distinctive rias (valleys flooded by sea-level rise following the last ice age), mountainous areas (which appear brown), and grasslands and peat bogs (green). At top and center right, snow covers the mountains of Scotland and North Wales, respectively.

An extratropical low-pressure system

PHOTOGRAPHS, SUCH AS THIS IMAGE OF A DEPRESSION taken during a space-shuttle mission, are not normally used for forecasting, but they provide useful information about the growth and decay of weather systems. This depression formed over the Gulf Stream near Cape Hatteras in North Carolina. Most of the spiral of cloud consists of an occluded front, where the cold front (bottom, nearest the camera) has caught up with the warm front (short section at extreme top right) and lifted the warm air away from the surface.

The smoke plume from Mount Etna

VOLCANIC PLUMES POSE A GREAT HAZARD for aircraft. Mount Etna (10,902 ft/ 3,323 m) in Sicily is Europe's largest and most active volcano: an orange lava flow is visible running south from the crater. On a clear day, such as here, when this image was obtained by the SPOT Earth–resources satellite, the plume is easy to track. When there is heavy cloud cover and complex weather patterns, determining the probable danger zone becomes far more difficult.

Atmospheric water vapor

THIS MAP FROM THE AQUA SATELLITE shows areas of low humidity around North and South America as light blue and those with high humidity, such as over the Caribbean, as dark blue. In the north, snow and ice appear yellow, but the small yellow patches in the tropics indicate individual thunderstorms with heavy precipitation. The dark stripe (right) shows loss of data over the Atlantic.

199

Water vapor over the world's oceans

AS WITH THE PRECEDING IMAGE, AREAS OF LOW HUMIDITY appear light blue, and those with high humidity (mainly in the tropics) are dark blue. Ice and snow over both the Arctic and Antarctic appear in yellowish tints, and some desert areas (particularly the Atacama Desert in South America and the deserts in Central Asia) have a slight yellow-green tinge. As the name implies, the Aqua satellite is specifically designed to monitor the Earth's hydrological cycle.

No Valid Data

1   2   3   4   5   6   7   8

Wave Height(m)   Sept 25 - Oct 4, 1992

Global map: mean wave heights

THE TOPEX/POSEIDON SATELLITE, orbiting 830 miles/1,336 km above the Earth, carried a radar altimeter to determine mean wave height across the globe, color-coded here from magenta (less than 3 ft/1 m), through blue, green, and yellow, to red (23–26 ft/7–8 m). The data was obtained at the end of September and the beginning of October 1992. The largest waves are in the south Indian Ocean and the Southern Ocean near Antarctica, and are typical of the season (southern winter/spring).

Pacific Ocean wind speed and direction

THE ARROWS REPRESENT WIND DIRECTION, and the colors wind speed, over the Pacific Ocean.
Blue shows wind speeds of 0–9 mph/0–14 kph, purple and pink speeds of 10–27 mph/15–43 kph,
red and orange speeds of 28–45 mph/44–72 kph. High wind speeds occur near a depression
located in the South Pacific (where the circulation is particularly marked) and also south of Alaska.
The distinct, narrow band with low wind speeds beneath the Intertropical Convergence Zone is
visible, running right across the Pacific just north of the equator.

Source of the Gulf Stream

SEA SURFACE TEMPERATURES AROUND THE FLORIDA PENINSULA are shown color coded, from gray and dark blue (0–35.6°F/0–2°C), through light blue, green, yellow, and orange, to red (72–79°F/22–26°C). Large-scale eddies are visible in the Gulf of Mexico at bottom left. The very warm Florida Current, passing between Florida and the Bahamas, is joined by the very broad, warm Antilles Current, flowing east of the Bahamas, to become the Gulf Stream. This powerful current flows north along the eastern seaboard of North America, before turning east off Cape Hatteras to become the North Atlantic Current.

Radar image of a squall line

GROUND-BASED RADAR PROVIDES INFORMATION about the intensity of rainfall and is often used in "nowcasting" (the preparation of short-term forecasts for no more than six hours ahead). The most advanced form (known as Doppler radar) also provides information about the direction and strength of the wind. Such details are vital in preparing warnings of imminent severe weather such as extreme hailstorms or tornadoes.

Ozone hole: Antarctic

THE GREATEST DESTRUCTION OF OZONE occurs in spring (in this case, October), when strong sunlight returns to the polar region after midwinter darkness. The Antarctic ozone hole is particularly strong, because a polar vortex of high-speed winds surrounds the pole, flowing essentially without obstruction right around the world. This cuts the polar region off from the general global circulation, preventing the replenishment of ozone from farther north.

Ozone hole: Arctic

IN THE ARCTIC. THE OZONE HOLE is much smaller than its Antarctic counterpart. This is primarily because the northern polar vortex is much weaker than the vortex in the south. The northern landmasses (and particularly mountain ranges such as the Rockies in North America) disrupt the circulation, and greater mixing occurs between polar air and that from lower latitudes.

209

Development of an El Niño event

COMPUTER REPRESENTATIONS of four stages of an El Niño event, showing conditions right across the Pacific. The map at top gives an idea of the relationship of the section to the Americas (top right) and Asia (top left). Sea temperatures vary from red (86°F/30°C) to dark blue (46.4°F/8°C). The height of the sea surface is shown in greatly exaggerated relief.

IN THE FIRST IMAGE (top left), taken in January 1997, there is a deep pool of warm water—and its associated rainfall—in the western Pacific (i.e., on the Australian and Asian side), where the ocean surface is actually higher than it is near the coast of South America. Near the latter, there is an upwelling of cold water, leading to highly productive fishing grounds. The thermocline (the division between warmed surface water and cold, deep oceanic water) slopes steeply toward the west. Such conditions correspond to the "normal" (non–El Niño) state.

IN THE SECOND IMAGE (top right) from June 1997, the surface winds driving the warm water to the west have weakened, and the pool of warm water is spreading eastward across the Pacific. Upwelling along the coast of South America has ceased, and a very thin, warm surface layer has risen. The slope of the thermocline has decreased dramatically.

IN THE THIRD IMAGE (bottom left), obtained in November 1997, the El Niño event is well under way. The deep pool of warm water and the accompanying area of heavy rainfall have nearly reached South America. The ocean surface is now higher in the east than in the west. Drier conditions (and an Australian drought) prevail in the western Pacific. The fishing industry off the coast of Peru has collapsed.

BY MARCH 1998 (bottom right), the pool of warm water has reached the South American coast. The associated, unaccustomed, torrential rain caused flash flooding, mudslides, and loss of life in Peru. Other events farther afield, particularly in the northern Pacific and in the Indian Ocean, are also linked to El Niño events.

211

Global sea temperature, July

IT IS PROXIMITY TO THE OCEAN and the temperature of its surface waters that largely determine the weather and climate occurring in any particular region. Ocean currents transport more heat from the tropics toward the poles than the atmospheric circulation. This computer model of sea surface temperatures is based on satellite data. The sea temperature varies from 95°F/35°C (yellow) in the tropics, through red, blue, purple, and green, to 28.4°F/-2°C (black) in the polar regions. Land appears gray.

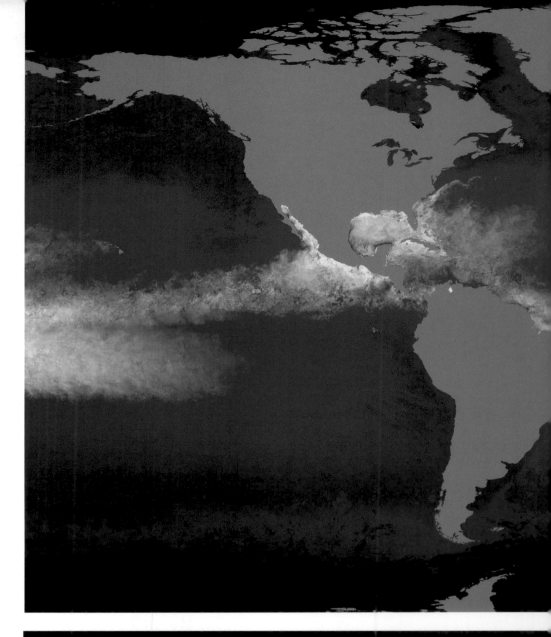

Global sea temperature, December

THE YELLOW TINT INDICATING the warmest sea surface temperature has shifted south in this computer-generated image derived from satellite data for December (during the southern hemisphere's summer). It is striking, however, that some regions of the North Atlantic off North Africa are now warmer than they were in July, where water warmed during the northern summer has spread across the ocean.

212

Previous pages | Atmospheric temperature anomalies

ANOMALIES ARE REGIONS where one of the atmosphere's properties (in this case temperature) varies significantly from the long-term average. Here, red areas indicate regions where the temperature is higher than the long-term average, and blue areas where it is lower. The red area over the Pacific, off the coast of South America, is typical of the pattern shown during El Niño events. This climatic phenomenon recurs roughly every ten years, causing major changes in ocean circulation, as described earlier, and severely affecting climate and biological productivity. Associated effects (known as teleconnections) influence the weather over most of the world, being strongest over the southern hemisphere and the northern Pacific.

Butterfly effect and chaos theory

FOLLOWING A DISCOVERY MADE while modeling weather systems on a computer, meteorologist Edward Lorentz published his paper, "Predictability: Does the Flap of a Butterfly's Wings in Brazil Set off a Tornado in Texas?" The idea, related to chaos theory, states that small events can make a massive difference further down the line, and it has been popularized as the "Butterfly effect." Lorentz's paper, however, showed that ultimately the question was inherently unanswerable. Errors in weather forecasting are inevitable because of inadequate global coverage, errors in the initial data, our uncertain knowledge of the physics involved, and unavoidable approximations in the equations used for computation. The Butterfly effect implies limits on predictability, not that small events necessarily have major consequences. The photograph shows a blue morpho butterfly of a type that inhabits the rainforests of northern South America.

A weak El Niño event

THIS IMAGE SHOWS SEA-SURFACE TEMPERATURES and wind directions in March 2003 from satellite data. Orange areas are hotter than normal and blue areas cooler. The tropical Pacific is now monitored continuously by satellites and the nearly seventy moored buoys in the Tropical Atmosphere-Ocean Array (TAO). The El Niño phenomenon is part of a larger system, known as the El Niño Southern Oscillation (ENSO), discovered in 1923, which links atmospheric pressure at Tahiti, in the central Pacific, to that over the Indian Ocean, as measured at Darwin in northern Australia. When one is high, the other is low. ENSO events are now fairly well understood, and it is possible to predict how they will evolve and affect conditions off South America and outside the immediate area. A strong El Niño brings heavy rains to southern California. The event shown here was weak, because the pool of warm water lay in mid-Pacific, not close to South America.

217

THE EARTH DISPLAYS AN ASTOUNDING VARIETY OF CLIMATE ZONES, from the icy polar deserts to the humid rainforests of the tropics. The climate at any location is influenced by many factors, the most important of which are the latitude, the proximity to the ocean and the sea surface temperature found there, the direction and strength of the prevailing wind, and the altitude.

The hot, humid air rising over the equatorial regions flows north and south at altitude and descends at the middle latitudes, producing the midlatitude high-pressure zones, where clear skies and high temperatures prevail. It is here that the world's major hot deserts are located, such as the Sahara and Arabian deserts in the northern hemisphere and the interior desert of Australia in the southern. Because of the Earth's distribution of land masses, most of the high-pressure zones in the southern hemisphere lie over the oceans, where there are large regions of low precipitation.

Air flows from the midlatitude highs, some back toward the equator, giving rise to the trade winds, and some toward the poles. Because of the rotation of the Earth, the air is diverted toward the right in the northern hemisphere and toward the left in the southern. This gives rise to two zones of westerly winds. In the south, because land masses are small, high, persistent winds occur in a broad belt above the Southern Ocean surrounding Antarctica.

In the north, the greater land masses and, in particular, the existence of major mountain chains, such as the

North American Rockies and the high Tibetan Plateau, lead to great variability in the location of the polar front, where warm air from the south encounters cold air flowing south from the Arctic. Despite this variability, certain high- and low-pressure areas tend to recur frequently and determine both the short-term weather and the longer-term climate of particular regions. Apart from the midlatitude highs (such as the Azores–Bermuda High over the Atlantic), high pressure over Siberia, Arctic Canada, and Greenland dominates winter conditions. Less persistent low-pressure areas such as the Aleutian and Icelandic Lows also play a considerable part.

The interiors of large land masses have a continental climate, with low annual precipitation, extremely cold temperatures in winter, and high temperatures in summer. Areas close to the ocean enjoy a milder, maritime climate, with greater precipitation and smaller temperature ranges. This is particularly marked on the western sides of the continents, especially in western Europe and on the west coast of North America, and also on the southern tip of South America.

Great seasonal variations occur in regions subject to monsoon winds, where there is a major shift in the direction of the prevailing wind with the season. In particular, cold air flowing from the Siberian High in winter produces northerly and northeasterly winds and dry conditions. As summer approaches, the high-pressure area collapses and the jet stream moves to the northern side of the Himalayas and the Tibetan Plateau, allowing southwesterly monsoon winds to bring humid air and heavy rains to southern Asia.

8 | CLIMATES OF THE WORLD

Edge of Greenland ice sheet

THE GREENLAND ICE SHEET is second only to the Antarctic ice sheet in size and covers about 708,111 sq. miles/1,834,000 sq. km, roughly 80 percent of Greenland. Bedrock is near sea level, but the northern dome reaches an altitude of over 9,843 ft/3,000 m, and the southern about 8,202 ft/2,500 m. The ice sheet is largely contained by coastal mountains, but gaps in the mountain ranges allow individual glaciers to reach the sea. In winter, the ice sheet becomes the center of the Greenland High, a high-pressure area from which frigid air streams out over the surrounding regions.

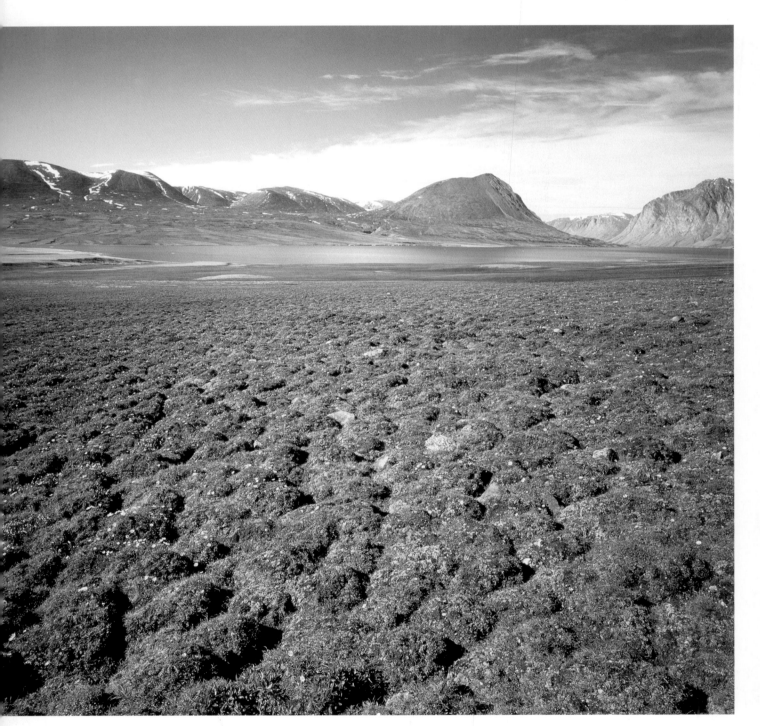

Hummocky ground in the tundra

THE WHOLE OF THE TUNDRA IS TREELESS, but certain areas may support mosses, lichens, and dwarf shrubs during the brief, warm summer season. The hummocks are a feature of much of the tundra and occur in areas of permafrost (permanently frozen subsoil). They (and the related areas of stones sorted into polygonal patterns, known as patterned ground) are created through the repeated freeze/thaw cycles, when larger stones migrate to the edge of the polygons. Any vegetation tends to hug the ground in the hollows between the hummocks, where it gains some slight shelter from the icy winds. The photograph shows Clavering Island in northeastern Greenland, at about latitude 74°N, where the pack ice melts in summer, as shown by the ice-free inlet in the distance.

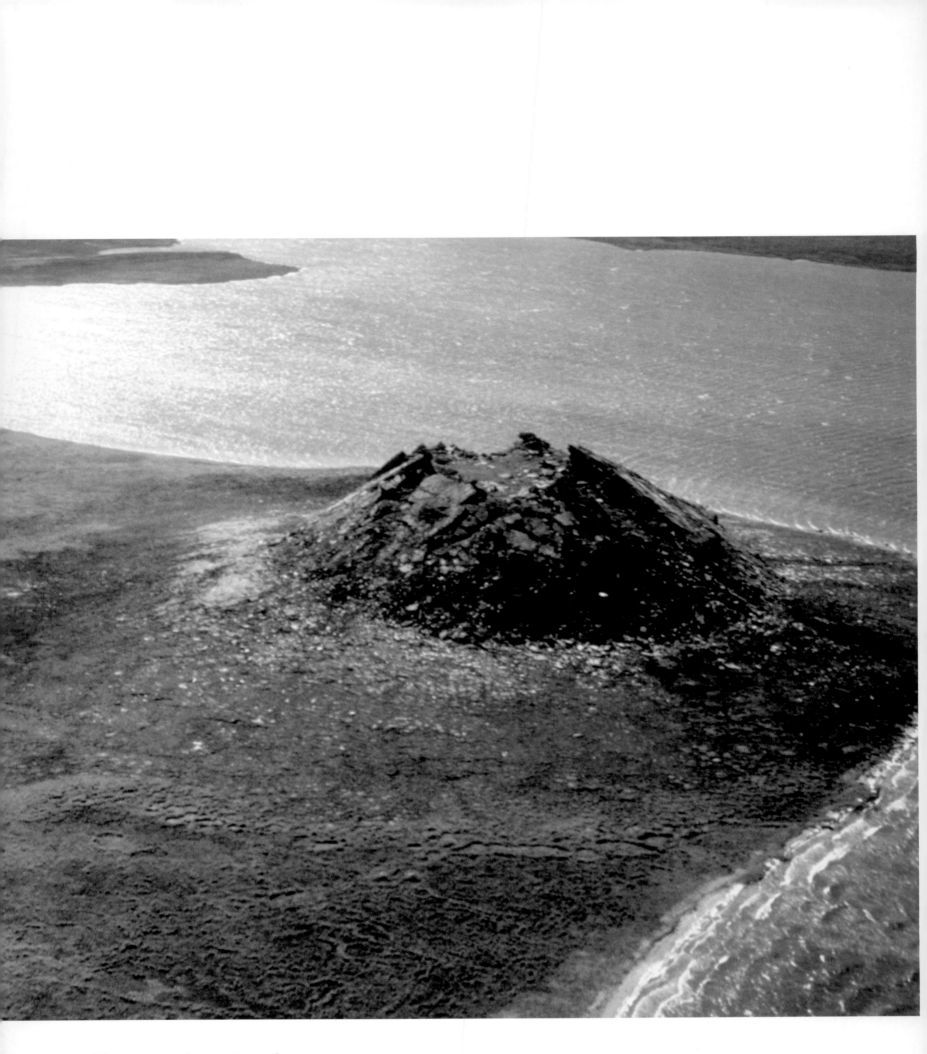

Pingo, northern Canada

THIS IMAGE SHOWS A PINGO IN THE PERMAFROST ZONE OF NORTHERN CANADA. Pingoes are highly characteristic features of permafrost regions. They occur in the swampy surface layer above the permafrost and consist of a lens-shaped core of ice, covered by a thin layer of earth or gravel. They gradually grow over a period of many years, and some have become substantial hills, reaching heights of 213 ft/65 m and diameters of 2,625 ft/800 m or more.

Pine forest in the taiga

THE TAIGA IS THE SUBARCTIC CLIMATIC ZONE that lies in an immense swathe across the northern hemisphere, between the treeless tundra and the boreal grasslands of Eurasia and the Canadian prairies. Overall, the zone experiences moderate amounts of rain, with extremely severe winters and peak summer temperatures of 59°F/15°C or less. It mainly consists of vast coniferous forests (as here in British Columbia, Canada) with a few very hardy deciduous species, notably birch. There is a distinct treeline on the mountains, above which trees are unable to grow, generally because of the decrease in temperature with altitude.

Temperate rainforest

THE CLIMAX VEGETATION (A NATURALLY BALANCED PLANT COMMUNITY) IN MARITIME west-coast climate zones (otherwise known as warm-temperate rain climates) is a temperate rainforest, seen here in the Olympic National Park, Washington State. This climatic zone generally has rainfall throughout the year, mild winters, and moderately cool summers. These conditions are found over a relatively restricted narrow band on the west coast of North America but over most of western Europe, where the climax vegetation is often deciduous, rather than coniferous, trees. Similar conditions occur in southern Chile, a small area around Johannesburg in South Africa, southeastern Australia, and New Zealand.

Mediterranean olive groves

THE MEDITERRANEAN CLIMATIC ZONE is a warm temperate one, with generally mild winters and moderate amounts of rain, and a dry season in summer. Around the Mediterranean itself, summer temperatures are generally above 71.6°F (22°C), but slightly lower summer temperatures are found in areas with a similar climate in California, South Africa, and southern Australia. These conditions are ideal for the cultivation of olives, vines, and citrus fruit.

The Florida Everglades

SIMILAR TEMPERATURE CONDITIONS to areas with a Mediterranean climate are found in the humid subtropical climatic zone but with a greater overall annual precipitation. In large regions of the United States (including the Everglades), Argentina, Brazil, China, southern Japan, and a small area of New South Wales in Australia, rainfall is high throughout the year. Northern India, Burma, and northern Vietnam have less rainfall in winter.

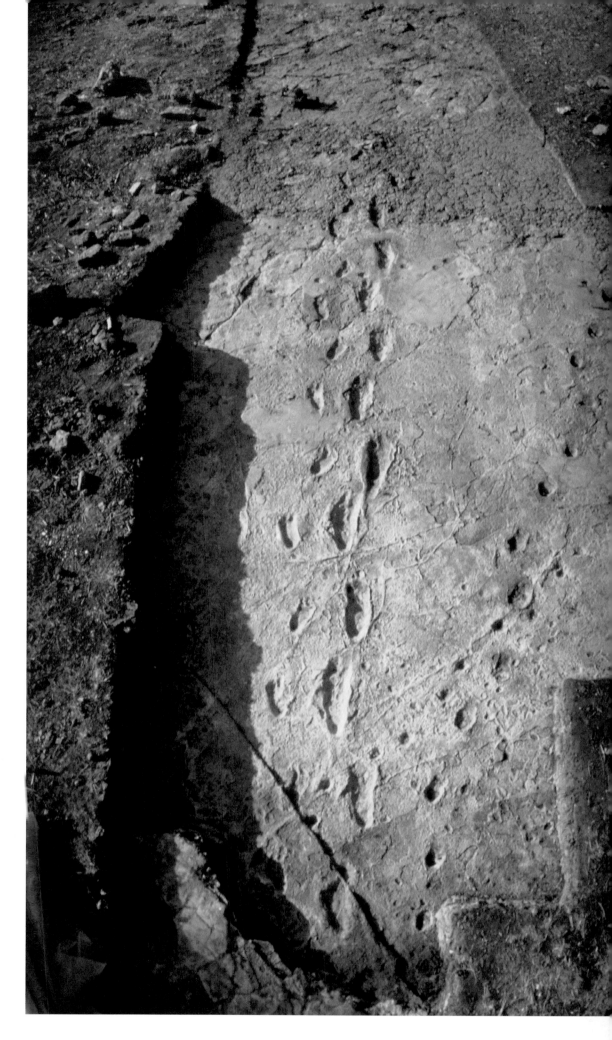

Ancient hominid footprints in the savanna

THIS TRAIL OF FOSSILIZED FOOTPRINTS in volcanic ash, at Laetoli in Tanzania, probably created by 237
Australopithecus afarensis, an early hominid, 3.6 million years ago. (Those on the far right were created by
a Hipparion, an extinct three-toed horse.) Early hominids may have developed a fully upright, bipedal gait
when climate change occurred in East Africa. Tropical rainforest gave way to humid, open grasslands of
the savanna, where an upright stance would be advantageous. Savanna regions have high temperatures
through the year and usually one major rainy season. They cover swathes of central Africa, South America
(particularly Brazil), and some regions of India, Indochina, northern Australia, and Central America.

Tropical rainforest, Ecuador

THE ORIENTE TROPICAL RAINFOREST in eastern Ecuador (left), lying east of the Andean Cordillera, is part of the great Amazonian rainforest. Tropical rainforests, which are primarily found here, in central Africa, and in Indonesia and New Guinea, experience considerable rainfall throughout the year. There is never less than 2.4 in/60 mm per month, although there is often a distinct rainy season, when rainfall may be much higher. The humidity, combined with the relatively constant, high temperatures, means that the vegetation continues to grow throughout the year, leading to an extremely rich and productive ecosystem with an incredible range of species of plants, insects, birds, and animals.

A cloud forest

CLOUD FOREST VEGETATION (RIGHT) develops at altitude in a few isolated areas in tropical regions. Here, low temperatures due to altitude and prevailing winds rising over the mountains bring almost continuous rain or mist throughout the year. This results in a forest consisting of a canopy of tall trees and a dense understory of shrubs and plants. The conditions are particularly suitable for epiphytic plants such as orchids and bromeliads, which flourish in upper canopies. This cloud forest lies at 7,218 ft/2,200 m on the eastern slopes of the Andes in Ecuador. There are signs that, with warmer temperatures, cloud forests are starting to shrink and that their lower boundaries are starting to retreat to higher altitudes.

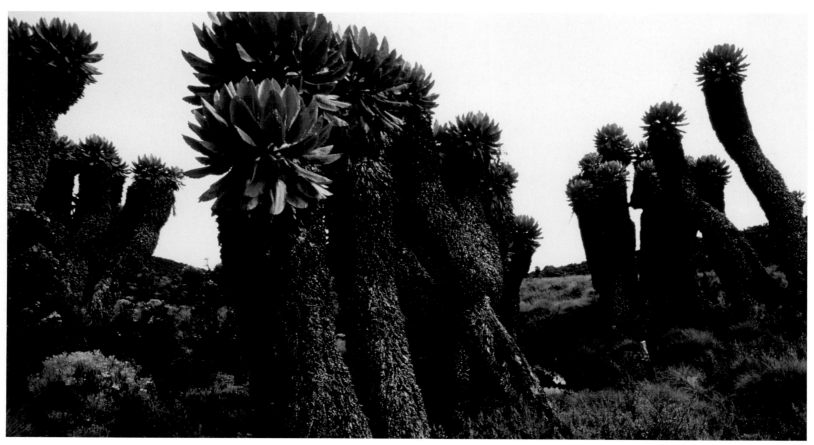

Giant groundsel on Mount Kilimanjaro

THE RELATIVELY ISOLATED, EQUATORIAL MOUNTAINS of Africa (Kilimanjaro, Elgon, and Kenya, in particular) have a high-altitude moorland zone, notable for giant groundsel and giant lobelia. These species have adapted to the low nighttime temperatures at the high altitudes. (In the groundsel, the rosette of leaves closes around the central bud at night to protect it from the frost.) This photograph was taken on Mount Kilimanjaro, Tanzania, at an elevation of 12,139 ft/3,700 m.

239

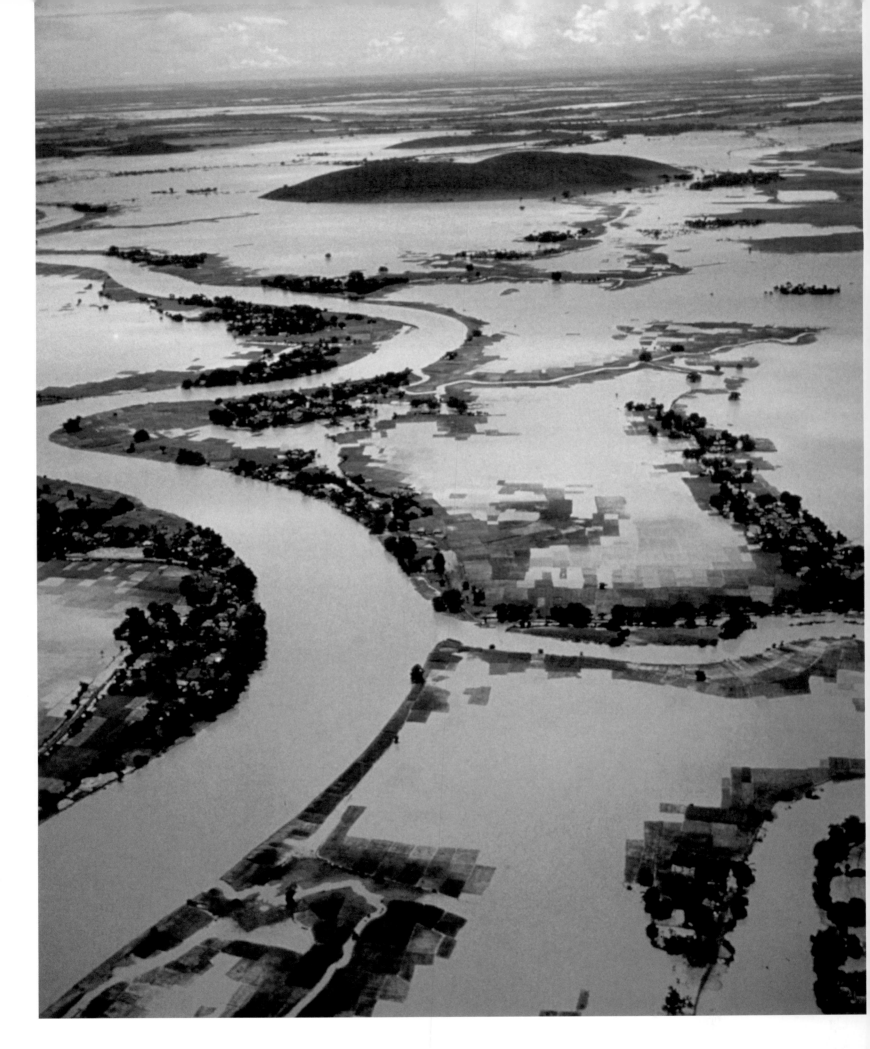

Monsoon flooding on the plains of northern India

MONSOON CLIMATES SHOW A MARKED REVERSAL between winter and summer wind directions, accompanied by a short dry season in winter and very heavy rains over the rest of the year. The most pronounced monsoon conditions apply over Asia, with particularly heavy summer rainfall in India, Bangladesh, Burma, Thailand, and the Philippines. Somewhat similar conditions occur in West Africa (in Sierra Leone and Liberia) and in small areas of South America (Guyana, Surinam, and French Guiana) and northern Australia.

Wind-sculpted rocks

YARDANGS ARE A CHARACTERISTIC FEATURE of desert or semidesert areas where there is little precipitation but strong, persistent winds. The soft, poorly consolidated rocks are eroded into ridges and gullies, parallel to the direction of the prevailing winds. The differences in height may be considerable, often reaching several meters. Yardangs are one of the few landforms created solely by the action of the wind, but they are found in many areas of the world. This yardang is in the Sahara, in Egypt's Western Desert. The most extensive fields occur in Central Asia, such as around Lop Nor, the dried-up salt-lake bed in the Sinkiang Autonomous Region of China.

The River Senegal and the Sahel

IN THIS FALSE-COLOR SATELLITE IMAGE, the River Senegal in West Africa, with its extensive flood-plain (gray), marks the boundary between the arid desert of Mauretania with its linear sand dunes (top) and the more humid savanna of Senegal (bottom). This boundary region is known as the Sahel, and some years ago there was considerable concern that the semiarid, steppelike lands of this region were expanding at the expense of the savanna farther south, perhaps as a result of overgrazing and deforestation in the Sahel region. It now appears, however, that there is a long-term, natural climatic cycle. The northern and southern boundaries of the Sahel move backward and forward, and the width of the Sahel itself expands and contracts over a period of decades.

Western Desert, Egypt

THIS COMPOSITE OF INFRARED- AND VISIBLE-BAND IMAGES shows part of Egypt's Western Desert, northeast of the Dakhla Oasis and on the edge of the vast Sahara Desert. Linear sand dunes lie parallel to the direction of the prevailing wind. The areas of exposed rock (brown) are scarred by dry river channels (known as wadis), cut during the rare flash floods that affect the region at long intervals.

Giant sand dunes, Namib Desert

THE DESERT IN NAMIBIA is one of the driest areas in the world. There is very little rainfall, because the Benguela current, which flows north along the coast, is extremely cold, suppressing evaporation. The little precipitation that occurs is primarily derived from fog, which forms when warm air is cooled by the cold surface waters and then drifts inland. The sand dunes in the Namib Desert are the largest found anywhere in the world, and some reach 1,148 ft/350 m in height.

Extensive wadis in coastal desert of Yemen

IN THE COASTAL DESERT OF YEMEN as in all deserts, the surface rocks are generally subjected to extreme temperature changes from day to night. The expansion and contraction fractures the rocks, leading to boulder-strewn areas and large expanses of rocks and gravel as well as the sand that is commonly thought to be characteristic of deserts. Any rainfall leads to flash floods, which rapidly, and deeply, erode the unvegetated surface and produce the characteristic wadis (dry river channels) and outwash fans of waterborne debris.

Overleaf | Salar de Uyuni, Bolivian Altiplano

THE ALTIPLANO (ALSO KNOWN AS THE PUNO) is a high-altitude desert in the Peruvian and Bolivian Andes. It consists of a series of basins, separated by mountainous ridges. In the north lies Lake Titicaca, and to the south Lake Poopó, the Salar de Coipasa, and the vast Salar de Uyuni. The latter is an enormous salt pan, the remnants of an ancient lake, about 4,093 sq miles/10,600 sq km, lying at an altitude of 11 995 ft/3,656 m. The polygonal pattern is reminiscent of the hummocks and patterned ground found in areas of the tundra or polar desert.

245

NEARLY ALL METEOROLOGISTS AND CLIMATOLOGISTS AGREE THAT GLOBAL WARMING IS TAKING PLACE, primarily through an increasing concentration of carbon dioxide in the atmosphere. It is exceptionally difficult to build suitable models to predict the sequence of events that is likely to occur and the rate of warming. By comparison, predicting the weather a few days ahead is a relatively simple task, because such predictions primarily involve just the atmosphere.

Predicting future climate is far more complex. Many additional factors must be taken into account. The oceanic circulation is exceptionally important in determining climate, because it transports great quantities of heat from the tropics to the poles. One fear is that global warming will cause a substantial portion of Greenland's ice cap to melt or calve as icebergs. This would introduce large quantities of freshwater to the North Atlantic, where, under current conditions, extensive evaporation causes the water to become highly saline and thus dense. The water sinks and then flows south along the oceanic floor, establishing a circulation that involves all the world's oceans. In the Southern Ocean, this bottom water is reinforced by frigid water sinking in the Weddell Sea, and the flow passes into the Indian and Pacific Oceans, eventually rising to the surface and returning via various branches to the North Atlantic. This circulation (known as the Great Ocean Conveyor Belt) is technically called a thermohaline circulation, because the flow is driven by differences in temperature and salinity. If this flow were to be interrupted, it would have major consequences for the climate of many areas of the world, most notably northern Europe.

In the atmosphere itself, the role of clouds is extremely important: the greater the cloud cover, the greater the amount of solar energy that is reflected back into space. Yet clouds also act as a blanket, preventing heat from the surface from escaping, so their effects are complex. With increased temperatures, there will be increased evaporation and increased cloud cover. A recent discovery is that there is evidence that rising temperatures have led to increased precipitation over East Antarctica and that this is acting as a temporary buffer against rising sea levels. Eventually, however, the Antarctic ice will begin to flow more rapidly into the sea, and sea levels will inevitably rise. It has also recently been established that meltwater is now penetrating to the base of the Greenland ice sheet, so that in summer large areas are no longer frozen to the bedrock, as had been thought, and are flowing seaward at a faster rate. This is of even more immediate concern.

The pollutants introduced into the atmosphere that produce extensive haze over large parts of the world also appear to have been buffering global warming. The "global dimming" that they have produced has reduced the amount of solar energy reaching the surface. The effect seems to be diminishing, however, so it has offered only a temporary respite from increased global temperatures.

9 | CLIMATE CHANGE

The Glen Canyon Dam

THIS DAM LIES ON THE COLORADO RIVER, on the Arizona–Utah border, where it has created Lake Powell (top). It is 709 ft/216 m high and one of the largest concrete dams in the world. The production of cement is one of the principal sources of carbon dioxide and thus a major contributor to global warming. For every ton of cement that is manufactured, one ton of carbon dioxide is produced. In the United States alone, about 500 million tons of concrete are used every year, with cement forming between 10 and 15 percent. That single source alone, therefore, releases between 50 and 75 million tons of carbon dioxide into the atmosphere every single year.

Previous pages | Icebergs calving from the Hubbard Glacier

MANY ICEBERGS ARE CREATED BY THE COLLAPSE of glaciers into the sea in the process known as calving. When the snout of a glacier reaches the sea and the ice starts to float, immense strains are placed on the terminal wall. The sections of ice that break away are the main source of large icebergs in the northern hemisphere, whereas sea ice is much thinner. The glaciers of Greenland, the northern Canadian islands, Spitzbergen, Novaya Zemlya, and (as here) Alaska are the main sources in the north. In the south, glaciers contribute icebergs from Patagonia and the Antarctic Peninsula, but giant tabular icebergs are also produced from the many ice shelves around Antarctica.

Concrete

CEMENT'S CONTRIBUTION TO GLOBAL WARMING does not end with its production. Quite apart from the indirect contribution made as it is transported long distances, it releases even more carbon dioxide when the chemical reactions occur that cause it to set. This colored environmental scanning electron micrograph (ESEM) shows gypsum crystals (brown) that have formed in setting concrete (blue). Concrete consists of cement, water, and aggregates such as sand and gravel. Adding water causes complex chemical changes to occur, hardening the cement around the aggregates, and it is at this point that the carbon dioxide is released.

251

Rice paddies and methane production

RICE PADDIES ARE A SURPRISING AGENT OF GLOBAL WARMING. This is not an abstract painting but an aerial view of extensive, terraced rice paddies in Richvale, California. The anaerobic (oxygen-deficient) conditions found in paddy fields (and other wetlands) mean that these are a major source of methane. (Other significant sources include ruminant animals and termites.) Although only traces are present in the atmosphere, its effect as a greenhouse gas is far greater than that of a similar quantity of carbon dioxide. Any increase in methane levels therefore has extremely serious consequences.

Termites and methane emissions

TERMITE MOUND IN MOSI O TUNYA, ZAMBIA. Termites are social insects that live in large colonies and build mounds that may be as tall as 23 ft/7 m and have a diameter of 98 ft/30 m. Termites are one of the few living organisms that are able to break down the cellulose and lignin in dead wood through the action of specialized bacteria in their digestive systems. In doing so, they release nutrients into the soil (and thus make a major contribution to the local ecosystem). Unfortunately, this process simultaneously releases large quantities of methane into the atmosphere.

1984   75.8 TG CH4

SOURCE: LERNER, MATTHEWS & FUNG (1988)   NASA/GISS

0   1   2   3   4   5   6   7   8   9   10+

1000 KG CH4/SQ KM/YR

Methane emissions by animals

THE METHANE PRODUCTION BY DOMESTIC ANIMALS is shown in this computer-generated map of the world. The map was created by applying the estimated annual production of methane by specific animals (cattle, dairy cows, water buffalo, sheep, camels, goats, pigs, horses, and caribou) to their population. The scale indicates metric tons of methane produced per square kilometer per year. (Black areas show a lack of data.)

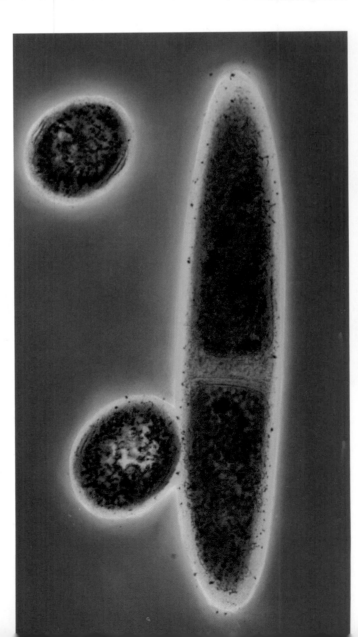

Methane and archaebacteria

ALTHOUGH CARBON DIOXIDE is the major component of global warming, methane (CH_4), is not only a potent greenhouse gas but also contributes to ozone depletion. Although there are some geological sources of methane, most is produced by extremely primitive organisms—more primitive than bacteria—known as archaea, here *Methanospirillum hungatii*. This is an anaerobic organism that lives in oxygen-free conditions and metabolizes carbon dioxide and hydrogen to produce methane. Unfortunately, the effect of reducing the amount of carbon dioxide in the atmosphere is far outweighed by the increased greenhouse effect from the newly formed methane.

254

Hydroelectric power and global warming

CONTRARY TO ITS POPULAR IMAGE, HYDROELECTRIC POWER is not entirely "green." Apart from the environmental and human costs of construction, hydroelectric reservoirs are a source of methane. The deep water is often anaerobic (oxygen-poor) and forms a suitable environment for methanogenic organisms. In addition, drawing off water in dry seasons and the subsequent refilling in rainy seasons means that the shorelines are also ideal habitats for methanogens. Reservoirs in tropical regions, such as this one at Catete in Angola, produce far greater quantities of methane than those in temperate zones where temperatures are much lower. The overall effect on global warming is often greater than a fossil-fuel power station producing the same amount of electricity.

CLIMATE CHANGE

Global atmospheric archive

ONE OF THE PROBLEMS FACING climatologists is that records of the concentrations of gases and pollutants in the atmosphere do not extend very far into the past. To try to remedy this situation—at least for the future—atmospheric samples, collected at the Cape Grim Baseline Air Pollution Station at the northwestern tip of Tasmania, are here seen stored in pressurized gas cylinders at the Global Atmospheric Sampling Laboratory (GASLAB) in Melbourne, Australia. The archive may be used to check concentrations of gases that were not monitored in the past or to refine existing measurements as analytical techniques improve.

Deep bore Vostok ice-core samples

CORES HAVE BEEN RECOVERED from a depth of 2,082 miles/ 3,350 m below Antarctica that seem to date to 400,000 years ago. Important information may be obtained about the prevailing temperatures throughout that period from determination of the ratios of oxygen isotopes that are present. Water molecules (H_2O) containing the lighter isotope, ^{16}O, are preferentially evaporated from the oceans, which are correspondingly enriched in ^{18}O. The Antarctic and Greenland ice caps, however, are enriched in ^{16}O and deficient in ^{18}O. The fluctuations in the ratio of the two isotopes are temperature dependent, enabling past temperatures to be determined. Worryingly, the cores show that abrupt, major changes in climate have occurred in just decades rather than over centuries as had been expected.

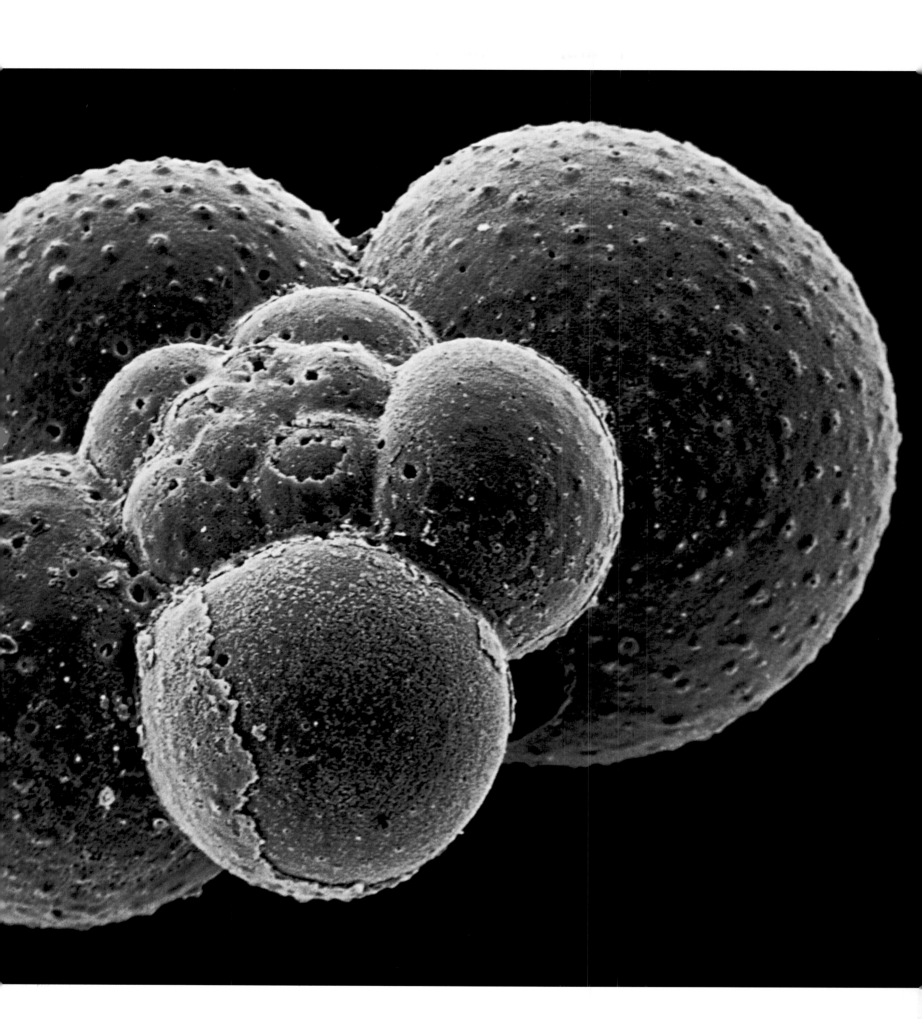

Marine foraminafera

FORAMINAFERA ARE SINGLE-CELLED ORGANISMS that are widespread throughout the oceans and are a valuable indicator of past climates. In some species that occupy multichambered shells, or "tests," the direction in which they coil is closely dependent upon the ambient temperature, altering at higher or lower values. Although this provides some information about prevailing temperatures, even more important information is derived from an analysis of the ratio of the two oxygen isotopes, ^{16}O and ^{18}O, in the calcium carbonate ($CaCO_3$) of tests recovered from seafloor cores. During ice ages, the polar ice caps contain particularly large quantities of ^{16}O, and foraminafera tests, formed from seawater, are greatly enriched in ^{18}O.

Air pollution, northern India, Bangladesh, and the Bay of Bengal

THE HAZE CONTAINS A MIXTURE of liquid and solid particles generated by industry, vehicle emissions, and domestic fires. Worldwide, such pollution is so widespread that, together with slash-and-burn agriculture and other sources of smoke and haze, it is largely responsible for "global dimming"—a reduction in the amount of sunlight that reached the Earth's surface in recent decades. This partially offset any effects of global warming, but the dimming now seems to be decreasing. Such pollution must be further reduced, but this is likely to lead to an increase (possibly dramatic) in global warming. Both issues must be tackled urgently and simultaneously.

Aircraft contrails over the Central Rhône Valley, France

THIS PHOTOGRAPH ILLUSTRATES ANOTHER FACTOR to be taken into account in modeling climate change. There are some cirrus clouds in the west (left), but most of the clouds consist of condensation trails (contrails), which form when the atmosphere is humid (in dry air, contrails evaporate quickly). Persistent contrails reflect sunlight back into space, cooling the surface below them. Climate models not only need to include realistic models for contrails (and all clouds) but also reliable estimates of future levels of air travel. In this image, Geneva and Lake Geneva in Switzerland are at the top center. The border between Switzerland and Italy lies between the two snow-free valleys at top right (the headwaters of the Rhône in the north and the Val d'Aosta in the south).

Saharan dust over the Atlantic

DUST FROM THE SAHARA HAS BEEN CARRIED AROUND the northern side of a depression by the surface winds and is affecting the Canary Islands (bottom center). Fine dust from the Sahara often spreads right across the Atlantic and makes a significant contribution to haze and particulate levels over North and South America. The dust actually provides valuable nutrients to the Amazonian rainforest. If global warming produces greater rainfall over the Sahara, with a return of the extensive vegetation that once covered the area, dust storms are likely to weaken, or cease completely, accelerating the decline of the rainforest on the other side of the ocean.

Beneath desert sands

THE BLACK-AND-WHITE STRIP IS A RADAR IMAGE of part of the Sahara Desert in southeast Egypt. It is superimposed on an image obtained at a visible wavelength. The radar pulses penetrate the surface deposits of dry, windblown sand and reveal features in the bedrock below, including drainage channels cut when the area had a much wetter climate. Such a technique is possible only over arid regions, because moisture in the soil reduces the depth to which the radio waves penetrate.

Ash plume, Mount Pinatubo

A FALSE-COLOR IMAGE, TAKEN ON JUNE 11, 1991, of the plume of volcanic ash from Mount Pinatubo on the island of Luzon in the Philippines. The yellow areas indicate normal clouds. Quite apart from the effects of ashfall, lava flows, and pyroclastic clouds, eruptions frequently exhibit some effect on the weather in the surrounding region. Areas under the plume suffer lower temperatures, and aerosols act as condensation nuclei, frequently leading to heavy rain, which can produce highly destructive mudflows (known as lahars) as the unconsolidated volcanic deposits are swept away by the rain. The area around Pinatubo suffered greatly from lahars when a tropical cyclone and its accompanying torrential rainfall moved in from the Pacific.

Plinian column, Mount Saint Helens

VOLCANOES THAT HAVE PLINIAN-TYPE ERUPTIONS eject vast quantities of material, much of it as a vertically directed column of gas, lava fragments, and pulverized rock. In this eruption of July 22, 1980 (after the initial catastrophic blast of 18 May), the column reached a height of about 7.5 miles/12 km. Other eruption columns, such as that of Mount Pinatubo in the Philippines in 1991, reach 12.4 miles/20 km or even more, sending aerosols (finely divided liquid droplets or solid particles) well into the stratosphere.

Caldera of Mount Tambora, Sumbawa Island, Indonesia

THE CALDERA (4 miles/6 km wide and 2,133 ft/650 m deep) was formed in 1815 by the greatest eruption of historic times, far greater than the better-known eruption of Krakatau in 1883. Aerosols from Tambora's eruption were ejected high into the stratosphere and were produced in such vast quantities that they not only affected the tropics but also blanketed the whole Earth. They caused a dramatic reduction in the amount of solar energy reaching the surface and had an immediate effect on the weather. The winter of 1815–1816 was exceptionally cold and persistent, and both spring and summer in 1816 were abnormally cold, so that 1816 became known as "the year without a summer." Crop failures and even famine conditions were widespread.

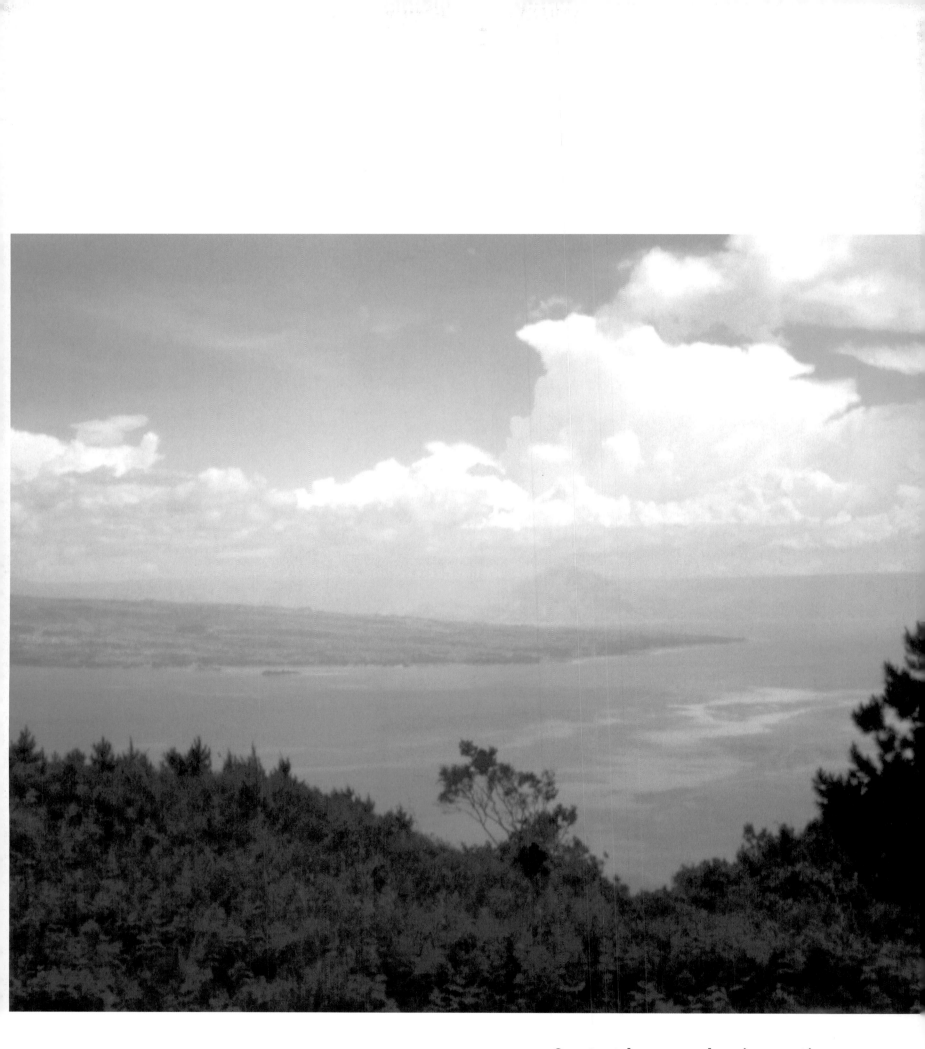

Greatest known volcanic eruption

THIS IMAGE SHOWS THE ENORMOUS CALDERA OF MOUNT TOBA IN SUMATRA, left by the greatest eruption known to volcanologists. Lake Toba occupies an area of about 502 sq. miles/1,300 sq. km, including Samosir Island, which is about 19 miles/30 km long by 6 miles/10 km wide. The gigantic eruption occurred in prehistoric times, estimated to have been around 74,000 BP (Before Present). The amount of ejecta is incredible: it appears to have been about 720 cubic miles/3,000 cubic km. For comparison, the volume of ejecta from Mount Tambora was about 12 cubic miles/50 cubic km, and from Mount Saint Helens a mere 0.7 cubic miles/3 cubic km. Such a gigantic eruption must have had a devastating effect on the Earth's weather.

GLOSSARY
&INDEX

CREDITS

Thunder Bay Press
An imprint of
the Advantage Publishers Group
5880 Oberlin Drive
San Diego, CA 92121-4794
www.thunderbaybooks.com

Text Copyright © 2006 Storm Dunlop
Design and Layout © 2006 Octopus Publishing Group Limited

Library of Congress Cataloging in-Publication Data available on request.

ISBN-13: 978-1-59223-700-5
ISBN-10: 1-59223-700-2

1 2 3 4 5 10 09 08 07 06

Commissioning Editor: Karen Dolan
Project Editor: Joanne Wilson
Editor: Katie Hewett
Design: Austin Taylor
Index: Sue Bosanko

Printed in China

Special thanks to Kevin Davis at the Science Photo Library.

Caption for pages 282–283

METEOROLOGISTS FROM THE SEVERE THUNDERSTORM Electrification and Precipitation Study (STEPS) launch a weather balloon into a storm in Kansas. This is a tornadic supercell thunderstorm, a massive storm in the process of developing a tornado. The weather balloon carries instruments to obtain data on temperature, pressure, wind speed, and electrical fields within the storm.